The Chemistry of Money

The Chemistry of Money

By

Brian Rohrig
USA
Email: brianrohrig@icloud.com

Print ISBN: 978-1-78262-983-2
EPUB ISBN: 978-1-78801-711-4

A catalogue record for this book is available from the British Library

The Royal Society of Chemistry is a charity, registered in England and Wales, Number 207890, and a company incorporated in England by Royal Charter (Registered No. RC000524), registered office: Burlington House, Piccadilly, London W1J 0BA, UK, Telephone: +44 (0) 20 7437 8656.

Visit our website at www.rsc.org/books

Printed in the United Kingdom by CPI Group (UK) Ltd, Croydon, CR0 4YY, UK

Preface

"A rich man is nothing but a poor man with money."

W.C. Fields

When I was a kid I found $10. I still vividly remember everything about that day. I was riding my bike to a park when, lo and behold, there was a $10 bill innocently lying along the side of the road, neatly folded in half. I couldn't believe my luck! When I arrived home, I breathlessly announced to anyone within earshot my good fortune. Even though this find happened over 40 years ago, it is still so vividly etched into my memory banks that it might have occurred yesterday.

We remember momentous, life-changing events—getting married, the birth of a child, landing a coveted job. Experiences packed with emotion stick with us, often for a lifetime. Many combat veterans never fully recover from the severe trauma they encountered during wartime. But positive emotional experiences are just as likely to stick with us as negative ones—if you score the winning goal in the championship game you will never forget it. I still remember winning a history award in school, although perhaps not as vividly as finding that $10 bill by the side of the road.

No matter how bohemian you might consider yourself to be, there is no denying the magnetic pull of money. At the

The Chemistry of Money
By Brian Rohrig
© Brian Rohrig 2021
Published by the Royal Society of Chemistry, www.rsc.org

Money Museum in Chicago the most popular display is a suit-case filled with one-million dollars. People gather around and stare with a kind of reverential awe at the stacks of $100 bills neatly arranged in the suitcase, 100 stacks in all. You can get your picture taken by it and receive the photo as a souvenir. In mine, I'm grinning like a Cheshire cat, thrilled to be so close to so much wealth.

In the prologue to his fine book *The Power of Gold: The History of an Obsession*, Peter L. Bernstein has this to say about gold, which of course has served as money throughout much of the world's history and in many ways still does today:

> This book tells the story of how people have become in-toxicated, obsessed, haunted, humbled, and exalted over pieces of metal called gold. Gold has motivated entire so-cieties, torn economies to shreds, determined the fate of kings and emperors, inspired the most beautiful works of art, provoked horrible acts by one people against another, and driven men to endure hardship in the hope of finding instant wealth and annihilating uncertainty. (Excerpt re-produced from ref. 1 with permission from John Wiley and Sons, Copyright 2001.)

Considering the incalculable toll of human misery brought on by mankind's quest for gold and all of its various incarnations, the Bible's declaration that the love of money is the root of all evil should come as no great surprise. Albert Einstein echoed these sentiments, saying, "Money only appeals to selfishness and always tempts its owners irresistibly to abuse it".[2]

Numerous thinkers have tried to shed some light on hu-mankind's near-universal mania over money. Freud postulated that money and feces were irrevocably intertwined. He argued that the universal identification of money as dirty or filthy is no accident. In "Character and Anal Erotism," Freud wrote:

> We know that the gold which the devil gives his paramours turns into excrement after his departure... Indeed, even according to ancient Babylonian doctrine, gold is "the feces of Hell..." It is possible that the contrast between the most precious substance known to men and the most worthless,

which they reject as waste matter ("refuse"), has led to this specific identification of gold with feces.

The original erotic interest in defecation is, as we know, destined to be extinguished in later years. In those years, the interest in money makes its appearance as a new interest which had been absent in childhood. This makes it easier for the earlier impulsion, which is in process of losing its aim, to be carried over to the newly emerging aim.[3]

With the advent of functional magnetic resonance imaging (fMRI) technology in the 1990s, for the first time it was possible to look inside the brain, so to speak, in an attempt to figure out what really makes us tick. We could find out what we really thought about money, and how it might alter our brain chemistry. During magnetic resonance imaging a strong magnetic field, aided by a radio frequency pulse, causes all the hydrogen atoms within the water molecules of the body to line up in the same direction. In fMRI, changes in blood flow to particular areas of the brain can be measured with pinpoint accuracy. Oxygenated blood responds differently to a magnetic field than deoxygenated blood, providing that subtle difference that the fMRI scan can detect. So, when you read about the latest study that relates how a certain area of the brain "lights up" when a test subject looks at pictures of puppies, it means that fresh blood is flowing into that specific area of the brain, presumably a different area than when that same subject looks at pictures of broccoli. There are six basic emotions—surprise, anger, disgust, fear, sadness and happiness. Different regions of the brain are activated whenever one of these emotions is experienced. So, in theory, an FMRI scan can tell us what emotion we are feeling when we respond to a certain stimulus.

A plethora of studies have shown that looking at money, handling money, thinking about money and expecting to receive some money all light up certain areas of the brain like a Christmas tree. One study showed that the pain processing centers of the brain were activated when subjects taxed with spending a limited amount of money were faced with unreasonably high prices.[4] Another study showed that when participants viewed money being destroyed, they registered a strong emotional response as evidenced by activation in the temporoparietal network.[5] When the pleasure centers of the brain are activated, a little shot of dopamine is released, reinforcing addictive behaviors like gambling, with compulsive gamblers receiving larger doses of this neurotransmitter than non-gamblers when both are exposed to the same stimulus.[6] One study revealed that the brain scans of those about to make money were strikingly similar to those who were doing cocaine.[7] Yet another study found that participants were better able to tolerate the pain of putting their fingers in hot (50 °C) water after handling money than were those who were handling pieces of paper that were simply shaped like money.[8]

The unifying theme behind these studies is that money elicits a deep emotional response within our brains. If my brain was being scanned when I found that $10 bill as a kid, it surely would have been all lit up, registering a response right up there with falling in love or anticipating my favorite meal. Money can make us feel good—for some it's their drug of choice—but like any drug it has its dark side. History is replete with examples of many mighty who have fallen, unable to reign in their primitive urge for money and the power and gratification it promises to bring.

While our brains might seem hardwired to respond to money in the same way it responds to food or sex or drugs, the physical object of these affections is by no means uniform across cultures. There is no universal conception regarding the physical manifestation of money, although gold comes pretty close. The entire notion of money is a social construct, subject to time and place.

While we will examine all aspects of money in this book, our primary focus will be the chemistry behind its physical form. The development of money did not occur in a vacuum, and as our knowledge of materials grew our money followed suit; indeed, today's currency bears only a superficial resemblance to its earliest forms. In many ways, the history of money parallels the history of science. You don't need a degree in chemistry to enjoy the journey on which we are about to embark, only a good dose of curiosity.

REFERENCES

1. P. L. Bernstein, *The Power of Gold: The History of an Obsession*, John Wiley & Sons, Inc., New York, 2000, p. 1.

2. A. Einstein, *The World As I See It*, Philosophical Library, New York, 1949, p. 25.
3. S. Freud, Character and Anal Erotism, *The Standard Edition of the Complete Psychological Works of Sigmund Freud*, Volume IX (1906–1908): Jensen's 'Gradiva' and Other Works, 1908, 173–174.
4. B. Knutson, S. Rick, G. E. Wimmer, D. Prelec and G. Loewenstein, Neural predictors of purchases, *Neuron*, 2007, **53**(1), 147–156.
5. C. Becchio, J. Skewes, T. E. Lund, U. Frith, C. Frith and A. Roepstorff, How the brain responds to the destruction of money, *J. Neurosci. Psychol. Econ.*, 2011, **4**(1), 1–10.
6. P. Anselme and M. J. F. Robinson, What motivates gambling behavior? Insight into dopamine's role, *Front. Behav. Neurosci.*, 2013, 7, 182.
7. H. C. Breiter, R. L. Gollub, R. M. Weisskoff, D. N. Kennedy, N. Makris, J. D. Berke, J. M. Goodman and S. E. Hyman, Acute effects of cocaine on human brain activity and emotion, *Neuron*, 1997, **19**(3), 591–611.
8. Z. Xinyue, D. V. Kathleen and F. B. Roy, The Symbolic Power of Money: Reminders of Money Alter Social Distress and Physical Pain, *Psychol. Sci.*, 2009, **20**(6), 700–706.

Acknowledgments

I owe a debt of gratitude to Professor Christopher Faulkner, author, historian, award-winning numismatist, and Professor Emeritus and Distinguished Research Professor at Carleton University in Ottawa, Canada. I would like to thank Professor Faulkner for lending his considerable expertise to this project, as his numerous comments and suggestions have led to a much richer work. His painstaking review of the manuscript and attention to detail saved me on multiple occasions. Any and all errors, however, are strictly my own.

The Chemistry of Money
By Brian Rohrig
© Brian Rohrig 2021
Published by the Royal Society of Chemistry, www.rsc.org

Contents

The Chemistry of Money
By Brian Rohrig
© Brian Rohrig 2021
Published by the Royal Society of Chemistry, www.rsc.org

Chapter 6
Counterfeiting **254**

CHAPTER 1

The History of Money

Gold makes the ugly beautiful.

Molière

Suppose you were unfortunate enough to find yourself in prison (falsely accused of course). Your wealth might very well depend on how many cigarettes you owned. In prison, cigarettes are like gold (or so I am told). Even if you didn't smoke you would want to amass as many as possible. Cigarettes are a valuable commodity that you could trade for something else you wanted, like a late-night snack or an extra dinner roll. With many prisons now outlawing tobacco products, a single cigarette can be worth $20 USD, and a pack $200 USD.[1]

If you were to visit a casino, your first order of business would be to trade in your money for casino chips, available in various denominations. Some casinos offer $100 000 USD chips for the really high rollers. The use of this artificial currency makes it very easy to place your bets. At the end of the night, you cash in your chips for real money, or if you have my luck, you leave with nothing.

If the earth were to suffer an apocalypse owing to a global catastrophe such as nuclear warfare or global warming, the most valuable commodity would not be gold or silver, but something rather more practical, maybe sunscreen if the ozone layer has been destroyed.

The Chemistry of Money
By Brian Rohrig
© Brian Rohrig 2021
Published by the Royal Society of Chemistry, www.rsc.org

1.1 EARLY FORMS OF MONEY

As each of the preceding scenarios illustrates, the medium of exchange in any transaction depends on what a particular culture deems precious at the time. Money can be loosely defined as anything of value that can be exchanged for something else. Throughout history, what societies have prized has changed considerably. In ancient times, it was not uncommon for workers to be paid with barley. Axe heads were commonly used as currency in early Mesopotamia. In Fiji, whales' teeth were once used as a medium of exchange. On the Indonesian island of Alor, bronze kettledrums at one time served as money. Up until World War II, even coconuts were used as currency in the Nicobar Islands of Southeast Asia.

Many of these early types of money were commodities, as they possessed value in and of themselves, as opposed to representing something else that has value. Commodity trading is still going on today. If you have a hundred bushels of wheat you possess something of value that can be traded for something else, and if the market crashes you can still eat, provided you know how to grind wheat into flour and then bake it into bread.

Another type of money is representative money, which represents something else of value. Paper money started out this way, and for the most part remained that way until the gold standard was abolished. Most money used today is fiat money, which only has value because the government says it does—it has no intrinsic value in and of itself, unless perhaps you were freezing to death and you needed something to burn.

Sometimes the lines between the different types of money are blurred. The ancient Aztecs used cacao (pronounced ka-KOW) beans as a source of currency. These beans could certainly be viewed as a commodity, but their societal value was much higher than their intrinsic worth would suggest. Cacao beans are the dried and fermented seeds from the *Theobroma cacao* tree, from which chocolate is made. Translating from the Greek, *Theobroma* means "food of the gods". The cacao tree can live for up to 200 years, but only produces quality beans for perhaps 25 years. The cocoa the Aztecs enjoyed was quite bitter and was served as a hot foamy drink, bearing little resemblance to the sugar-laden chocolate products we enjoy today. The bitterness is due to the alkaloid compounds in the cacao beans, such as theobromine, as well as polyphenols, which actually make the biggest contribution to the bitterness of chocolate, as just a little can go a long way. A polyphenol, as its name implies, is a compound composed of a large number of phenol molecules. Phenol consists of a hydroxyl group (–OH) connected to a benzene ring. The concentration of polyphenols is reduced markedly during the processing of the cacao bean after harvesting, making the final product more palatable.

The Aztecs used cacao beans for a lot more than making hot chocolate—these beans were their predominant form of money, a practice that began with the Mayans. They even had an elaborate accounting system to keep track of their wealth—the beans would typically be stored in bags containing 8000 beans, just like today's coins that come from the mint bagged in predetermined quantities. The Aztec emperor Montezuma II, the wealthiest of all the Aztec emperors, had a billion cacao beans in storage, most of which was exacted as tribute from the vanquished tribes of his vast empire. To gain an idea of the enormity of this sum, consider that in those days an avocado would set you back three cacao beans. The last time I bought an avocado it cost $1.79 USD, putting the value of a cacao bean at about 60 cents—a respectable sum. The Spanish even had established monetary exchange rates with the Central American natives—in 1535 140 cacao beans could be exchanged for 1 Spanish real.

I recently purchased a one-pound bag (454 g) of raw cacao beans (also known as nibs) for $12.99 USD; there were 151 beans in all, which works out to about 9 cents per bean. Considering all that has transpired over the past 500 years, I would say the cacao bean has held its value quite well.

However, not everyone appreciated the value of the cacao bean. In the16th century the English often enlisted the aid of pirates to counter the Spanish navy. These pirates had little understanding of the value of these precious beans, and if they captured a ship that was carrying them they would dump them overboard. In 1579 a band of English pirates boarded a Spanish ship looking for gold. Instead, all they found was a ship full of cacao beans, which they mistook for animal droppings—they promptly burned the ship and its cargo, destroying a vast fortune.

When one considers the tremendous purchasing power of cacao beans, it is easy to appreciate the exotic nature of such a drink, and why it would likely be consumed only on special occasions, as you would literally be drinking your money. It would be somewhat like grinding up your precious metals and consuming them, which is exactly what happens at some high-end restaurants. If you're ever in Dubai, you will want to stop by *Gold On 27*, a bar on the 27th floor of the Burj Al Arab Jumeirah hotel. One of their most famous cocktails is the Element 79, which contains flakes of pure gold, making its $40 USD price tag seem almost like a steal. It is non-alcoholic, as consuming pure gold should be intoxicating enough. Gold, by the way, has no taste.

Believe it or not, cacao beans were so valuable they were actually counterfeited. Unscrupulous folks would fill a hollow shell with soil and try to pawn it off as the real thing, or they would try to pass off similarly sized objects as real beans. A seed from another plant could easily be dyed to look like a cacao bean. A trained eye and a deft touch though could be enough to spot a fake, so a wise merchant would take their time when making transactions.

Using seeds for money might appear a bit odd, but if you consider their chemistry it makes a lot more sense. When cacao beans are harvested, they are first removed from their pods. They are surrounded by a white layer of pulp, which is actually quite sweet owing to the high sugar content (mostly fructose and glucose). These sugars undergo fermentation by natural yeasts, converting the sugars into ethanol. This ethanol is then oxidized to acetic acid by bacteria. All of these reactions tend to be exothermic, raising the temperature of the cacao beans. The fermentation process and subsequent reactions impart the beans with their chocolaty flavor and color, and typically take around 5 days, after which the beans are dried out.

After drying, the moisture content will be very low, making the beans especially resistant to rotting. If the cacao beans are not dried out they will quickly be overtaken with mold, rendering them useless. Microorganisms love moisture, and without it most cannot thrive. Any type of foodstuff with a long shelf-life will have a low moisture content. Freeze-dried foods are especially long lasting.

When considering moisture content, food scientists are mostly concerned with available moisture (abbreviated Aw, for water activity), not necessarily the total percentage of water in a sample. The available moisture is rather the unbound or free water within a substance that is available to microorganisms, and can range anywhere from 0 to 1.0. Most microorganisms cannot grow in a food if the available moisture is below 0.60. A piece of bread has a water activity value of 0.96, while an orange is close to 0.99, helping to explain their susceptibility to mold. Crackers come in at 0.1 and instant coffee 0.168. A typical dried cacao bean will have a water activity level between 0.42–0.57—if the moisture level goes down much below this level the beans are susceptible to cracking. After processing the moisture content goes down even further—the milk chocolate we eat has a water activity level between 0.37 and 0.5,[2] making it an exceptionally mold-resistant food. I for one have never seen mold grow on a piece of chocolate, most likely because my sweet tooth virtually guarantees its rapid consumption.

Speaking of mold, I am reminded of a buckeye tree we had in our yard when our kids were little. The buckeye is an especially beautiful nut, smooth and shiny with an alluring chestnut brown color. One year we had an especially bountiful harvest of buckeyes, and my kids gathered up a huge bin full of them. The bin of buckeyes was eventually forgotten and abandoned to the basement. A few months later they had all grown moldy and we had to throw them out, making it easy to see why the buckeye has gained the reputation of being a worthless nut. Not only are they poisonous to people, but even animals avoid them. Buckeyes would not make good currency—their shelf life is simply too short. Substances with a longer shelf life tend to have the greatest value.

As the dried cacao bean so aptly demonstrates, in order to serve as money, an object must be stable, or in chemical parlance, inert. Cacao beans are not as inert as gold and silver, of course, but still their value was considerable. The cacao beans used by the Aztecs were only valuable because they were the end result of a time consuming, labor intensive process that would render them stable. They weren't just plucked from a tree. There is no value in an untreated, moldy cacao bean. However, once fermented and dried, they could last for up to a year.

Cacao beans are about 50% fat, by weight. The majority of these fats are saturated, making them especially stable. Although saturated fats are normally associated with animal products, they can be found in plants. A fat is considered saturated if every carbon atom within its fatty acid molecules is single-bonded, allowing for the maximum number of hydrogen atoms—the molecules are "saturated" with hydrogen. Butter, which contains a lot of saturated fat, can be left out for days on the kitchen counter with no apparent rancidity, owing to the stability of its saturated fatty acids. Owing to its stability, at one time the Irish even used butter as currency.

Cacao beans are not much to look at—they are not shiny like coffee beans, but are duller and somewhat shriveled in appearance. They are more reddish than brown, looking somewhat like a miniature desiccated sweet potato. Cacao beans can come in a wide assortment of colors, depending on the variety of tree from which they came, ranging from white to deep red. Recently ruby chocolate was introduced with great fanfare—it was the first type of natural chocolate to be introduced since white chocolate, which made its debut 80 years ago. Ruby chocolate does not contain any dyes or additives but is simply made from ruby-colored cocoa beans. Chocolate made from these beans is a pleasing pink.

The high fat content contributes to the relatively low density of the cacao bean. They are less dense than water, with an average density of 0.59 g cm^{-3}. You can easily verify this value by adding some beans to water and carefully measuring the percentage of the floating bean that is underwater—exactly 59% of the bean will be immersed.

Even though they taste bitter, the cacao bean actually has an acidic pH, which can be confirmed by adding a few drops of universal indicator to some ground up beans that have been added to distilled water. The solution will turn reddish orange, indicating a pH of around 5.5. The acidic pH is due to leftover acids from the fermentation process, namely acetic and lactic acids. If the beans are dried too quickly then too much acid will remain in the bean, producing a sour taste. A pH below 5.0 is generally considered unacceptable. If the pH goes too high, however, then the growth of mold is not inhibited, which is an even bigger problem.

An examination of other objects that served as money in earlier cultures reveals a similar tendency toward inertness. At one time Roman soldiers were paid partly in salt. The words *salary* and *salt* are both derived from *sals*, the Latin word for salt. Salad also comes from this same root word. Salads do not contain salt but salad dressing does; if you examine the label on a bottle of dressing you will typically see salt listed as one of the ingredients, contributing to the 200–300 mg of sodium per serving. Salami also hails from this same root word.

Salt never goes bad, being made of positive sodium ions linked up with negative chloride ions by incredibly strong ionic bonds. It is quite stable. The driving force behind chemical bonding is this tendency toward stability, or lower energy. Pure sodium will explode if placed in water. Inhaling pure chlorine gas, even in relatively small amounts, can be fatal. However, when joined together both of these unstable compounds will behave themselves and form a compound that is quite stable. It is even hard to melt—once I tried to melt some salt with a Meker burner, and even though the burner was supposed to easily exceed the 800 °C melting temperature of sodium chloride, the solid crystal lattice refused to give way.

So, if you are feeling underpaid at your current place of employment you can take solace in the fact you are not being paid in salt. At today's salt prices, the average worker would be taking home about a ton of salt per week.

Salt is not the only condiment that has served as currency. Believe it or not, pepper once served as money. Pepper comes from grinding up the fruits, known as peppercorns, of the black pepper plant (*Pepper nigrum*). Pepper has been a hot commodity since the days of the Romans, and even today, pepper is the king of spices, being the most widely used spice in the world (cinnamon comes in a close second). During the Middle Ages, pepper was worth more than its weight in gold!

To discover all the things that have been used for money one only has to look at the many slang words used to describe it. Have you ever referred to money as bread or dough? How about bringing home the bacon (and frying it up in a pan)? In the US, the word *buck* is still used to refer to a dollar. Early European settlers to North America would trade in buckskins, the hides of male deer. Although it would seem logical that as many females as males would be harvested, for some reason the term *doe* never caught on as a monetary unit.

Another slang term for money is *clam*. "Seashells," which are the discarded shells of mollusks, have been used extensively for money across the world, as recently as the late 19th century. Every time you talk about "shelling out" money for a purchase, you are paying homage to your ancestors, who had to use actual shells. One of the more popular varieties of monetary mollusks was the cowry shell, a sea snail common in the Pacific and Indian Ocean regions. The Chinese were probably the first to use cowry shells as money, perhaps as early as 300 BC. However, they were used for a lot more than money—these shells were also believed to carry magical properties, being commonly used by witch doctors and pagan priests. Owing to their resemblance to female genitalia, they were believed to possess the ability to increase fertility—if a woman wore a belt of cowry shells she would be more likely to conceive (Figure 1.1).

As with any shell, its primary ingredient is calcium carbonate ($CaCO_3$). Calcium carbonate can also be found in limestone, marble, chalk, eggshells and antacid tablets such as Tums. So, if you are suffering from acid indigestion you can always take a bite of chalk if Tums are not available, as chemically the two are identical. Calcium carbonate is very stable but its Achilles' heel is acid. Any introductory geology student knows the acid test for limestone: place a few drops of hydrochloric acid on a chunk of rock and see if it bubbles—if it does, calcium carbonate is likely present. The equation is as follows:

$$CaCO_{3(s)} + HCl_{(aq)} \rightarrow CaCl_{2(aq)} + H_2O_{(l)} + CO_{2(g)}$$

Figure 1.1   Cowrie shell.
          © Shutterstock.

This reaction helps to explain why the calcium carbonate skeletons of the coral reefs are being dissolved as the world's oceans increasingly become more acidic. Any acid will dissolve calcium carbonate, even weak ones, but the primary culprit in the oceans is carbonic acid. As more carbon dioxide is pumped into the atmosphere, more carbonic acid is produced, as the following equation shows:

$$CO_{2(g)} + H_2O_{(l)} \rightarrow H_2CO_{3(aq)}$$

Native American tribes would string together shells into intricately woven decorative mats to make wampum, which would eventually be adopted by European settlers as a way of exchanging goods with the Indians (Figure 1.2). In 1641 the colony of Massachusetts legally declared six wampum beads to be equal in value to one penny. The original intent of wampum, however, was ceremonial, in which it was revered as an art form. It is commonly reported that the island of Manhattan was sold to the Dutch by the Native Americans for 24 dollars' worth of wampum; many historians, however, doubt the veracity of this claim.

There is seemingly no end to monetary terms that derive their meaning from objects once used as money. The word *chattel* refers to any type of property besides real estate, and historically has been used to refer to larger items of value that a person might possess, such as cattle. The words *cattle* and *chattel* are both derived from the Latin word *capital*, referring to net assets.

Figure 1.2 Wampum beads.
© iStock/Getty Images Plus.

As slavery was commonplace throughout much of the world's history, chattel was commonly used to refer to slaves as well as cattle.

Along these same lines, the term pecuniary is derived from the Latin word *pecunia* which means money, which in turn is based on *pecus*, the Latin word for cattle. From this root word is derived the word peculiar, which originally referred to one's own distinctive property. The peccary, a large pig-like mammal from Central America, provides yet another derivation of this same word. Peccaries, as well as ordinary domestic pigs, have been commonly used as currency. The pig is still a major player in the global economy, as at any given time there are around a billion domestic pigs in the world.

1.2 THE STONE MONEY OF YAP

With fiat currency, objects used for money don't have to be useful as long as everyone agrees on their value. The inhabitants of Yap, a series of tiny islands in the Pacific Ocean, are famous for carrying the idea of fiat money to its most logical extreme. At one time their choice of currency was limestone rocks, which were imported from the island of Palau 300 miles away. Islanders called this stone money *rai*.

It was not an easy task to transport these rocks, and the bigger the rock the more perilous the journey required to get the rock to Yap. Therefore, the bigger the rock the more it was worth. They were usually carved to resemble large wheels, with a hole in the middle, looking somewhat like a giant bagel (Figure 1.3). They were transported by inserting a log through the center, after which they were rolled like a giant wheel. Some of the bigger rocks were as much as 4 meters in diameter, while the smaller ones were no larger than an Oreo cookie. The larger rocks were often left in place, as the ownership of the rock was common knowledge. One large rock being transported to Yap sank to the bottom of the ocean before it could be brought to shore, yet its owner was given credit for having obtained it, and he could still use it to engage in transactions.

The rai stones were mostly made from calcite, a crystalline form of limestone with a density of around 2.7 $g\,cm^{-3}$. The larger stones were around 4000 kg in mass, with a volume of around 1480 liters—about half the size of a sports car. On Moh's scale of

Figure 1.3　Yap rock.
© Hal Beral/Corbis/Getty Images.

hardness, calcite comes in at 3, making it relatively easy to carve and mold. The rocks were not only used as currency but were also considered to be works of art. Value was based not just on the rock's size, but also on its carvings and its luster.

The tribal chiefs controlled the circulation of the rai stones, setting a limit on how many could be imported to the island, thus ensuring that their value would remain high. These rocks could be used to buy food or land, and were even used to cement alliances between neighboring tribes.

The last stone was carved in 1931, and even though today the dollar is the official currency of Yap, the stones remain. You can visit the island of Yap today and see these stones. The surface of these once shiny rocks is now pitted owing to hundreds of years of exposure to the elements, with acidic rainfall being largely to blame. You don't have to sail to Yap to see these rocks, though— they are on display at museums around the world, including the Smithsonian in Washington D.C. and the Museum of the National Bank in Belgium.

It would be easy to castigate the inhabitants of Yap as foolish; after all, of what value is a giant rock? Yet, when you consider what most of us receive at the end of the week as remuneration for our labors, are we really all that different from the Yapese? The pieces of paper and shiny metals we receive are no more valuable than giant rocks, although they are more convenient. In addition, with the advent of the direct deposit, online bill paying and credit cards, we don't even see "real" money most of the time. However, we place great value on these particular items because our government has decreed that they are the medium of exchange that can be used to purchase other items we need or want.

1.3   THE RISE OF METALS

As society progressed and more and more people moved to cities the methods of monetary exchange had to adapt. Hauling around wheat and herding pigs was not feasible, and rolling around giant rocks made even less sense. However, the rocks of Yap can teach us a valuable lesson: they only achieved their elevated status because they were both unusual and rare. A garden variety stone pulled from the dirt obviously wouldn't cut it as currency, but a rarer one just might. But precious stones are just too rare—there simply aren't enough of them to go around. It was only a matter of time before metals pulled from the earth would become the primary source of the world's currency. As early as 2500 BC, gold, silver and copper were being traded, long before they were ever fashioned into coins. The history of money as we know it today parallels the history of man's attempts to extract these metals from the earth.

Metals were so essential to the development of civilization that the eras of ancient societies were named after the types of metals their people fashioned, beginning with the decidedly non-metallic Stone Age to the Bronze Age and finally to the Iron Age.

The Bronze Age ranged from around 3700 BC to 500 BC, depending on which part of the world one refers. The Bronze Age gave way to the Iron Age, which would last another 1000 years.

Before the Bronze Age gold, silver and copper were the predominant native metals generally known and used by man. Any naturally occurring metal, either in pure form or as an alloy, can be referred to as a native metal. These native metals—gold, silver and copper—have been dubbed the coinage metals, and were the principal metals used to make coins up until the middle of the 20th century. A brief examination of the chemistry of the coinage metals will serve to demonstrate why they were the material of choice to make coins for so many years.

1.4 THE ALLURE OF GOLD

Even today gold holds a special allure; however, it is not the most expensive element, by far. Most of the radioactive elements are more expensive than gold, but they are rather hard to come by, as most are produced in multimillion-dollar particle accelerators. Californium-252, for example, can run to a cool $27 000 000 USD per gram. There are at least six stable elements more expensive than gold, among them osmium, palladium, thulium, platinum, rhodium and lutetium. Of these, the rare earth metal lutetium is the most expensive, going for around $1000 USD per gram.

Despite their exorbitant cost, none of these more expensive elements can compete with gold for the affections of mankind. Gold has been the centerpiece of the world's monetary system throughout history, and in many ways still is today. At the time of writing, gold was selling for around $1213 USD per ounce ($39 USD per gram). In the marketplace gold is still sold by the ounce, despite the conversion to the metric system by every country in the world, with the United States, Myanmar and Liberia being the only holdouts.

Gold is not measured using the familiar avoirdupois ounce, however, which is equal to 28.35 g—16 avoirdupois ounces equals one pound, or 453.6 g. Instead, all precious metals are measured using the troy ounce, which derives its name from the city of Troyes, France, in which the unit first originated. There are only 12 ounces in a troy pound, with each troy ounce weighing 31.1 g— therefore a troy pound only weighs 373.2 g. Armed with this knowledge, you can add a new twist to the classic riddle when you ask which weighs more, a pound of feathers or a pound of gold.

Gold has been found on every continent and is even a component of seawater, with concentrations in the oceans estimated to be from 5–50 parts per trillion; even though this concentration doesn't sound like much, there is far more gold in the oceans than has ever been mined from the earth. Numerous attempts have been made to find an economical way to extract this dissolved gold, but so far, no method has been successful. The difficulty lies in the enormous task of going through so much seawater to get at it. As each liter of ocean water contains only a few nanograms of gold, the cost to extract it is far more than the gold is worth.

However, most gold is buried deep within the earth, with geologic processes working it up to the surface a little at a time. Gold is often associated with veins of quartz, but not always. Owing to its high density, it will tend to settle along the bottoms of creeks and streams. Panning for gold only works because gold is denser than almost everything else, sinking to the bottom of the prospector's pan.

You don't need to sift through ocean water or dig in the dirt to obtain gold, however, nor do you need to be an alchemist, as it can be synthesized in the lab. In 1980, Glenn Seaborg, the namesake of seaborgium (element 106), led a team at the Lawrence Berkeley National Laboratory at the University of California, Berkeley, USA, in which they bombarded bismuth foil with high energy beams of carbon and neon nuclei traveling at almost the speed of light. These high intensity collisions succeeded in knocking off four protons, as well as a number of neutrons, from each of the several thousand bismuth atoms, finally fulfilling the alchemist's dream of transmuting base elements into gold.[3]

The amount of gold produced was so small it could only be detected by measuring the radiation emitted by the unstable gold atoms that formed. There are actually 36 known radioactive isotopes of gold, with half-lives ranging from 186 days (Au-195) to 17 μs (Au-171). A few stable gold atoms (Au-197) were probably formed as well, but they were present in such small amounts that not even a mass spectrometer could detect them. As for the cost, Seaborg estimated that it would cost about $1 000 000 000 000 USD per ounce to produce gold by this method.

It is gold's scarcity that has contributed to its value. If all of the gold that has ever been extracted from the earth were fashioned into a single cube, each side would only measure a little more than 20 meters. At a density of 19.3 $g\,cm^{-3}$, an 8000 cubic meter gold cube would have a mass of 174 000 metric tons, with a value of over 6 trillion dollars! All of the money in the world today would add up to more than 80 trillion dollars, so unless a lot more gold is discovered, backing up money with gold is no longer possible.

Part of gold's appeal is its color. No other metal has a yellow color; the only other yellow element is sulfur, a non-metal. Gold's color makes it easy to spot, as it can sometimes be found glittering in streams or along banks. Its chemical symbol, Au, is derived from the Latin word for gold—*aurum*—which roughly translates into "shining dawn". It also refers to Aurora, the Roman goddess of dawn.

The observed color of any object represents the wavelengths of light that are reflected by that object. Most metals are silvery and shiny-looking because they reflect all wavelengths of light. We generally think of white objects as being the most reflective, but it is actually shiny objects that have the greatest reflectivity. Although gold has reddish hues, for simplicity's sake let's consider it to be yellow. Yellow, being a secondary color, is formed by the combination of red and green light. Therefore, when visible light strikes gold, blue wavelengths of light are absorbed and mostly red and green are reflected. This blue light absorption occurs when high-energy photons associated with blue light (200–300 nm) strike gold, exciting some of its 5d electrons up to the 6s orbital. The amount of energy required for this electron excitation is equivalent in energy to that of a blue photon, thus blue light is absorbed, and what we see is the reflected red and green light, which our brain perceives as yellow.

This electron excitation only occurs because electrons are loosely held in metals. These loosely held electrons comprise a sea of negatively charged particles that move freely between adjacent metal atoms. These delocalized electrons generally consist of the valence electrons—the electrons in the outermost energy level. As they are farthest from the nucleus, they are the most loosely held.

Most metals appear silvery in color because it takes an invisible photon, namely an ultraviolet (UV) photon, to cause the electrons to jump from one energy level to the next. However, if a UV photon is absorbed it does not change the color we perceive because UV is invisible to us. However, it takes less energy to excite gold's electrons, requiring only a blue photon as opposed to an ultraviolet photon. If it took more energy than this to excite an electron of gold then gold would not have its golden color, and we would presumably be calling it something else.

To explain why it only takes a blue photon rather than an ultraviolet one to excite gold's electrons, it is necessary to resort to Einstein's theory of relativity. Einstein claimed that many things we consider to be constant are actually not constant at all.

As velocity increases time slows down and mass gets bigger. The fact that everything is relative—even things like mass and time—is perhaps the most mind-blowing aspect of Einstein's theory. As gold's nucleus is so massive—79 protons—its electron cloud is pulled in very tightly. This trend of decreasing atomic radius is evident as one moves from left to right across any period on the periodic table. In period 6, for example, where gold is located, each element in this period has 6 major energy levels. Yet, the nucleus gains a proton for each successive atom in the period, and as a result the electron cloud is pulled more tightly inward as one moves across the period, resulting in the relatively large diameter of the alkali metals on the far left side of the table, and the very small diameter of the halogens on the far right side of the table. Gold, being near the far right side of the transition metals, has an especially small diameter for a metal.

It is gold's massive nucleus that effectively pulls in the electron cloud. In order to avoid being pulled into the nucleus, the electrons must move at extremely high speeds. The 1s electrons move at about half the speed of light, increasing their mass by about 20%. This increase in mass in turn causes the electron cloud to become even smaller than it normally would. Energy is conserved if these now more massive electrons are forced to traverse a smaller distance in their journeys. If you were twirling a ball on a string and you began moving it so fast that the mass of the ball actually began to increase, you would compensate by making smaller circles with the ball—it would simply require too much energy to do otherwise.

As a result of this more massive electron cloud being pulled increasingly inward, a smaller gap is produced between the 5d and 6s orbitals, thus requiring less energy for an electron to make the jump from a 5d to a 6s orbital. As a result, blue light is absorbed as opposed to more energetic ultraviolet light, causing mostly yellow light to be reflected. So, the next time you admire the golden luster of gold, you can be reminded that Einstein was right after all.

However, one could argue that color is not an intrinsic property of a substance but rather a perception in our brain. In the dark a piece of gold has no color, but metals also appear black if you grind them up into a fine powder. It's not that the particles are too small to reflect light—they are plenty big enough to do that or you wouldn't be able to see them at all—but rather in powdered form there exists numerous surfaces from which light bounces off. As light waves bounce around among particles a little bit of light is absorbed each time; if it happens enough all of the incipient light will eventually be trapped. If no light is reflected from an object it will appear to be black. If the concentration of gold in the earth falls below 30 ppm it is invisible to the naked eye. In most gold mines the gold itself is not even visible. Gold mining today is mostly about the pursuit of this invisible gold, with the most profitable gold mines today having to treat tons of rock just to get a few grams of pure gold.

For evidence of the "black color" of gold in its powdered form, try this little experiment: take a gold ring, or any piece of jewelry with gold in it and rub it across your face, pressing down firmly. If you are a male, chances are nothing will happen, but if you are a female, there is a good chance that a black mark will form on your face. However, there is an equal likelihood that the mark will *not* appear. Younger women with light complexions—typically blondes and redheads—tend to be most susceptible to receiving the black mark. The medical term for this effect is black skin dermographism, which literally means "black skin writing".

One hypothesis that has been formulated to explain this strange phenomenon is that a few gold particles rub off owing to abrasive particles in makeup, which females are more likely to wear than males. Titanium dioxide (TiO_2), for example, is commonly used in makeup and skin powders, and with a hardness of 5.5 on Moh's scale it is much harder than gold's 2.5–3. However, the black mark can appear when no makeup is worn, and may not appear even when makeup has been applied. One urban legend speculates that the presence of a black mark indicates iron deficiency; there is no evidence to back up this claim, however, which likely arose owing to the higher incidence of iron deficiency during a woman's menstrual cycle. The appearance of the black mark is still a medical mystery, and is probably brought about by an unknown combination of factors affecting skin chemistry, such as pH, sweat and hormones. Regardless of the reason, it does nicely show that gold in powdered form will tend to be black.

Tiny particles of gold are not always black—they can actually exhibit a variety of hues. Nanosized particles of gold can be pink, purple or red. At around 20 nanometers gold particles appear to be red; at 100 nm they look blue. Gold particles of various sizes have long been used to impart color to stained glass windows.

Gold's small size affects more than just its color—it also affects its density. The easiest movie bloopers to spot are those where thieves are hoisting bricks of gold as easily as they would ordinary bricks. In gold storage facilities, like Federal Reserve banks in the US, workers often wear protective metal covers made of magnesium over their shoes for protection when handling the very dense gold bars. Gold's "heaviness" is one of its most defining attributes.

Gold's abnormally high density is not only the result of the atoms themselves being so compact, but it is also due to how these atoms are arranged. Gold typically forms face-centered cubic crystals, having an atom not just at the corners of the cube, but on each of its faces as well. This arrangement allows for the maximum number of gold atoms to be crammed together into a very small space (Figure 1.4). Many other metals share this same crystalline structure, including silver and copper.

If you are fortunate enough to find a rather large piece of gold that has not been subject to erosion, you can observe its crystalline structure. Although not often observed, all metals can form crystals under the right conditions. As a crystal contains a

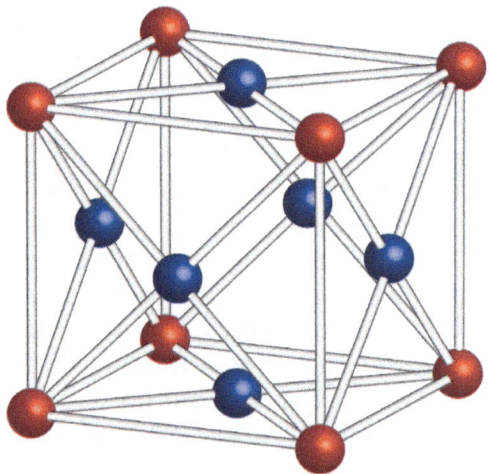

Figure 1.4   Face-centered cubic structure.
              © Shutterstock.

regularly repeating latticework of atoms arranged in a specific geometric shape, the longer it is allowed to form the bigger it will be. For an analogy, consider a group of people walking into an empty room, each holding a chair. If this group had only 5 seconds to arrange the chairs into neat rows, the result would be rather haphazard, but if they were given 5 minutes to perform this same task, they could arrange the chairs into nice neat rows. By the same token, crystals that are given a long time to form will be much larger.

1.5 ALL THAT GLITTERS IS NOT GOLD

While not many pure elements are yellow, there are numerous yellow compounds. The aptly named "fool's gold" has led to many false claims and no small amount of consternation among gold prospectors. Pyrite, or iron sulfide (FeS_2), has a cubic crystalline arrangement similar to gold's, but it differs in that it forms a simple cubic structure, having ions only at the corners of its latticework, making it much less dense than gold. The individual iron and sulfur atoms that make up the latticework also have much smaller masses than that of gold. Gold is almost four times as dense as pyrite, which comes in at a mere 5.0 $g\,cm^{-3}$. Incidentally, pyrite's specific crystalline structure was one of the first to be discovered through X-ray diffraction.

Pyrite differs from gold not only in density but also in color, being a darker yellow. Pure gold is not only lighter but also, well, more golden. If placed side by side, the color alone would be enough to make a distinction between the two. If rubbed across a porcelain plate, gold will leave a yellow streak whereas pyrite's streak will be black. Pyrite is much harder than gold, coming in at 6–6.5 on Moh's scale of hardness. A quick and easy way to test for the authenticity of gold in the field is to bite down on it, as tooth enamel has a hardness of 5, enabling you to leave a dental impression on a piece of gold. But don't bite down too hard—a piece of fool's gold could easily break your tooth! Gold is also much more malleable than pyrite; if hit with a hammer gold flattens out, while pyrite, being brittle, will shatter if struck.

Unlike gold, pyrite can be made in any high school chemistry lab. All you need to do is mix sulfur and iron in roughly equal proportions in a test tube and heat strongly over a Bunsen burner flame for about 20 minutes. The end product will not have a discernible crystalline structure, however, as it forms very quickly. It will also look mostly black, but if you look closely you can see some yellow highlights. You will know you have done a good job if your pyrite is not attracted to a magnet, as iron loses its magnetic properties when it bonds with sulfur.

Even though pyrite has been the bane of prospectors, it can actually lead to real gold. Just as veins of quartz will sometimes lead to gold deposits, so can deposits of pyrite. However, pyrite is much more common than gold, so the presence of one does not necessarily indicate the presence of the other. Pyrite may even contain small quantities of pure gold embedded within its crystalline structure; usually not much, though, perhaps 100 ppm. However, some of the largest gold mines in the world get their gold from pyrite. The most famous of these was discovered in 1962 near Carlin, Nevada, in the US—these types of gold mines are now called Carlin-type deposits, and can be found all over the world.

Removal of the gold from the pyrite is a heavy-duty industrial process using a lot of nasty substances—it is not a task for the weekend hobbyist. It typically involves the process of leaching, in which gold-containing ore is added to a liquid substance that reacts with gold but not the other impurities. Cyanide has traditionally been used for this purpose. Gold will react with cyanide to form the complex ion aurocyanide $[Au(CN)_2]^-$, leaving behind the undesirables. A complex ion is essentially a molecule with a charge. It has a metal ion at its center, to which is bonded other molecules or ions (called ligands) by a special type of covalent bond known as the coordinate covalent bond—a bond in which all the electrons are donated by only one of the atoms in the bond. (You can always recognize a complex ion by the brackets.) Ligands don't have to be anything special—they can be simply neutral molecules such as water, or ions like chloride or sulfide.

An additional process is then used to liberate the gold from the gold-containing complex ion. The biggest drawback with this method is that a lot of cyanide waste is produced, which is not the least bit friendly to the environment.

Other chemicals used to extract gold from its ores are just as nefarious, with mercury historically being the material of choice.

Gold will quickly and instantly dissolve in mercury, forming an amalgam—a mercury-containing alloy. I once witnessed first-hand mercury's affinity for gold when a package was delivered to the school where I was teaching. It was a mercury barometer that had been ordered by the previous instructor, which had been on backorder but had now arrived. Unbeknownst to me at the time, some of the mercury had leaked out of the barometer and was all over the packaging. After unpacking the barometer, I noticed that my gold wedding band had turned white. Once I realized what had occurred I immediately took it off, terrified that I might acquire mad hatter's disease or something worse. A chemist friend told me to soak the ring in a dilute solution of ammonium hydroxide, which would react with the mercury and restore the gold ring to its original condition. It worked like a charm.

Owing to fears about mercury's toxicity, safer alternatives have been sought to isolate gold. Researchers at Northwestern University in the US have had some initial success using cornstarch, of all things, as a way to separate gold from its ore.[4] They discovered that α-cyclodextrin (a cornstarch-derived polysaccharide composed of six glucose units), could be used to form a gold-containing precipitate when added to a solution of $KAuBr_4$. This precipitate can then be easily reduced to isolate the pure gold. The hope is that this environmentally friendly alternative will eventually allow cyanide to be phased out completely as a method of extracting gold from its ore.

Another method of promise for acquiring pure gold involves using gold-eating bacteria. It has been known for quite some time that certain species of bacteria thrive on heavy metal compounds that would kill most other living things, most notably *Cupriavidus metallidurans*. These rod-shaped bacteria aren't just immune to the deleterious effects of these noxious chemicals but rather seem to thrive in the types of toxic soups that would kill most other life forms. The secret to survival is specialized enzymes that can break down these highly toxic heavy metal compounds. They actually require copper, as their name implies, but only in trace amounts. In their search for nutrients they will gobble up any of a variety of heavy metal compounds, among them various gold complexes.

However, these bacteria have no use for gold, so whatever gold they happen to ingest in ionic form is excreted in atomic form.

They have the ability to reduce both Au^+ and Au^{3+} ions within these complexes into elemental gold.[5] Now golden bacterial droppings are quite tiny, being only a few micrometers in size, but when enough gold within these droppings gets together metallic bonding takes over and larger-sized particles form that can be visible to the naked eye. It is believed that much of the so-called secondary gold, or newer gold, in the earth's crust is due to the action of these gold excreting bacteria. Maybe Freud was right after all.

1.6   GOLD: THE MOST NOBLE METAL

Although gold's initial allure may have been owing to its color, its usefulness is a result of its stability. Gold is the most noble of the "noble" metals, a loosely defined group of heavy metals that tend to resist corrosion. As discussed earlier, gold has a very tiny diameter. Gold's tiny diameter is a big reason for its inertness—gold is the least reactive metal on the periodic table. Smaller metals are less reactive, as the reactivity of a metal is determined by its ability to lose electrons. Small metal atoms hold their electrons more tightly than larger metal atoms, thus forming stronger metallic bonds with their neighbors, which makes them less available as partners for other atoms that are looking to bond with them. Gold is, for the most part, immune to oxidation.

Gold does form compounds, but it is much more common to find gold in nature in its pure state than as an ore. Its reluctance to react has made it the element of choice for jewelry, coins and other ornamentation. Gold has been used in dentistry[6] for thousands of years, as gold is impervious to decay. Gold is even resistant to most acids. It will dissolve in a 1:3 mixture of nitric and hydrochloric acid, known as aqua regia, but will not dissolve in either of these acids by themselves. As most other metals are susceptible to either of these acids, one way to separate gold from other metallic impurities is by adding a strong acid.

The ability of gold to dissolve in aqua regia was used by two Nobel prize winning physicists to keep their gold medals out of the hands of the Nazis during World War II. Max von Laue had won the Nobel prize in Physics in 1914 for his work with X-ray diffraction in crystals. Although German born, Laue was an outspoken opponent of the Nazi party. Fearing that the Nazis would confiscate his gold Nobel medal, he entrusted it to Niels Bohr at his institute in Copenhagen, before Denmark fell to the Germans.

James Franke won the Nobel prize in 1925 for discovering the laws that govern the impact of an electron upon an atom. Being of Jewish heritage, he left Germany in 1933, shortly after the Nazis came to power. He too entrusted his Nobel medal to Niels Bohr for safekeeping. When the Nazis invaded Denmark, Georgy de Hevesy, a chemist working in Bohr's lab, was desperate to prevent these metals from being confiscated. De Hevesy was of Jewish heritage himself and had likewise fled Germany to avoid persecution. As it was a crime to export gold out of Germany, de Hevesy knew that everyone who worked at the lab could be implicated if the medals were discovered. As the Nazi invaders were marching through the streets, de Hevesy hurriedly dissolved the two medals in aqua regia, storing the orange solution in a flask on a shelf in his lab, which the Nazis never touched. Returning to the lab after the war had ended, de Hevesy was astonished to find the flask with the bright orange solution in the same spot that he had left it three years earlier. He precipitated out the gold and returned it to the Nobel Foundation in Stockholm. They recreated the medals using the original gold and presented them again to Laue and Frank in 1952.

Gold is considered to be among the most malleable and ductile of all the elements, owing to the delocalization of its electrons. A single gram of gold can be pounded into a one square meter sheet only a couple of hundred atoms thick, or drawn into a wire longer than the length of a football field. Its metallic pliability has allowed it to be fashioned into all sorts of useful items, from plates and cups to masks and jewelry, and eventually into coins. The ancients discovered that applying just a very thin layer of gold over an object would produce the same effect as a thicker layer. They became quite adept at capitalizing on gold's extreme malleability by developing a process known as gilding. The ancients were famous for gilding everything they could get their hands on. Today, the wealthy and privileged are often referred to as the gilded class. Gold's melting point is 1064 °C, much lower than copper or iron, making it relatively easy to melt and mold.

Far from being just pretty to look at, gold has many practical uses. When the first transatlantic cable system (TAT-1) was laid at the bottom of the ocean between Scotland and Newfoundland in 1955–56, corrosion of sensitive components by saltwater was a main concern. The twin copper cables were themselves deemed safe, as they were wrapped in various layers of plastic and tape. However, the repeaters in the cable, which contained amplifiers that helped to carry the signal, were a bigger concern. The solution was to plate many of these components with gold to protect them from corrosion. Within each repeater was a number of capacitors—their casing was plated inside and out with gold to protect them from corrosion. Gold was chosen because it did not form what was termed at the time metal "whiskers," crystalline growths that could short-circuit the electronic components of the repeaters.[7] The ease with which gold could be easily soldered made it especially desirable. The gold did its job well, as the TAT-1 served admirably for two decades before being retired in 1978.

If you take your cell phone apart, chances are you will see some gold—its excellent conductivity and resistance to corrosion make it ideally suited for electrical contacts. As gold is such a good conductor it is a terrible insulator, so static charge is less likely to build up in your phone. A typical cell phone contains about 50 mg of gold. That doesn't sound like much, but when you consider that worldwide a billion smartphones are purchased every year, that amounts to 50 000 kg, or 50 metric tons of gold. According to the World Gold Council, gold mining produces about 2500–3000 metric tons of "new" gold every year, so for now there's enough gold being produced to meet the demand. A more pressing problem facing the smart phone industry is the scarcity of the rare earth metals—your phone is full of them, and as their name implies, they are not always easy to come by. It's not that the rare earth elements are all that rare, but rather can only be found in small quantities in any single location, forcing engineers to be ever more creative if they run out of, say, praseodymium.

Gold is not that hard to find in your cell phone—if you see something gold colored it's probably gold, as there is not a lot of fool's gold lurking in there. The easiest place to see gold is on the circuit board. If you drop it in a strong acid, you can recover the gold, as the other metal parts will dissolve.

NASA has made extensive use of gold in the space program. Gold is an excellent reflector of both infrared and ultraviolet radiation, serving to prevent spacecraft from overheating, as well as protecting delicate electronic equipment. Spacecraft are often coated with gold foil, but gold isn't just a reflector, it also absorbs a fair amount of visible light. Even though the visor on an astronaut's space helmet is coated with gold he can still see because enough light will pass through. Other metals are much better reflectors of visible light; silver, for example, is the most reflective of all metals, making it a poor choice for use on spacecraft—if the sun hits it just right an astronaut would be blinded by the glare. As gold is not as reflective of visible light, it is a much better choice.

Gold even has medical uses. Its high density has been put to practical use by implanting bits of gold in the upper eyelids of those suffering from lagophthlamos, a condition in which the eye does not shut completely. The implanted gold adds just enough weight to enable the eyelid to shut all the way. This condition may not sound like such a big deal, until one stops to consider that every blink of the eye bathes the cornea in protective lubricating fluid. Without regular blinking, the eyes become dry and irritated. The problem is especially exacerbated at night, when sleeping with your eyes open leads to severe dryness, not to mention being somewhat unsettling to anyone watching you sleep. Gold is an effective treatment because of its high density—its large mass to volume ratio makes it hardly visible, yet it adds just enough weight to cause the eyelid to completely shut.

1.7 SILVER: GOLD'S FIRST COUSIN

Like gold, silver is visually appealing, owing to both its color and its luster. The chemical symbol for silver is Ag, which is based on the Latin word *argentum*, meaning shiny. Silver has long been associated with money, as it shares many properties with gold that make it especially useful as a means of monetary exchange. The French word for money is *argent*, and the Latin word for a banker that deals in silver is *argentarius.*

Silver was first mined around 3000 BC in Anatolia (present day Turkey). Silver has always played second fiddle to gold. First of all, it is not nearly as expensive. At the time of writing, a gram of silver was valued at \$0.54 USD per gram, 70 times less expensive than gold. Scarcity tends to determine cost, and silver is much more common in the earth's crust than gold; however, it is not as likely as gold to be found in its elemental form as it is more reactive than gold. Even though silver is subject to tarnishing, it is still less reactive than a lot of metals; fortunately, the tarnish is fairly easy to remove (see Chapter 4). Although you can occasionally find pure silver in nature, it is more common to find it as an ore, or alloyed with gold in a mixture known as electrum.

Silver owes its nonreactivity to its filled d orbitals. One of its valence electrons, which should be happily ensconced within a 5s orbital, instead is found in a 4d orbital, not only completely filling the d subshell, but also the entire energy level, conferring upon silver a great degree of stability. Copper and gold also share this tendency, making all three group 11 elements perfectly suited for use as coinage metals. Even though these elements occupy the 9th column of the transition metals, their electron configurations more closely resemble the elements in the 10th column of this block. Their electron configurations are as follows:

Copper: [Ar] $4s^1$ $3d^{10}$
Silver: [Kr] $5s^1$ $4d^{10}$
Gold: [Xe] $4f^{14}$ $6s^1$ $5d^1$

A filled subshell, and especially a filled energy level, is very stable—electrons will always arrange themselves in such a way as to attain the lowest possible energy state, and thus the greatest stability.

One way to separate silver from gold is to take advantage of silver's greater reactivity. Gold will not react with nitric acid but silver will, forming copious amounts of toxic reddish-brown nitric oxide (NO) gas in the process (not to be confused with laughing gas, which is nitrous oxide (N_2O)). Inhaling nitric oxide will most assuredly *not* make you laugh. The reaction is as follows:

$$3Ag_{(s)} + 4HNO_{3(aq)} \rightarrow 3AgNO_{3(aq)} + 2H_2O_{(l)} + NO_{(g)}$$

Silver is often found alongside lead. Lead ore, or galena (PbS), often contains silver. Unlike gold, silver combines much more readily with other elements to form compounds. Some common silver ores are argentite (Ag_2S) and chlorargyrite ($AgCl$), as sulfur and chlorine are frequent partners with silver. In addition, silver ores can consist of borates, chlorates, oxides, iodates, bromates, carbonates, and nitrates.

The most efficient way to remove silver from its ore is through the process of smelting. Smelting is not the same thing as melting, although a lot of melting often occurs during smelting. Whereas melting is simply a physical change, smelting involves the separation of a compound into its constituent elements. The ancients were quite adept at extracting pure metals from their ores. The first smelters likely stumbled upon their discovery by accident, perhaps by sifting through the ashes of a kiln or pit used to fire pottery, where they likely noticed some type of shiny metal that wasn't there previously. Smelting is typically a two-step process. First the ore must be heated, and then an additional substance must be added which reacts with the impurities through the process of reduction, leaving behind pure silver.

For example, suppose you have some silver oxide (Ag_2O), and you want to remove the silver from it. The silver must first be reduced, as its charge must go down from +1 in the compound to 0 in the pure element. Reduction always involves the addition of a reducing agent—carbon in the form of charcoal will often do the trick, which yields carbon monoxide upon heating. A typical reduction reaction might be:

$$2Ag_2O_{(s)} + C_{(s)} \rightarrow Ag_{(s)} + CO_{2(g)}$$

Silver sulfides require an additional step. First the ore is heated strongly in an atmosphere of excess oxygen, allowing oxygen to take the place of sulfur, according to the following reaction:

$$2Ag_2S_{(s)} + 3O_{2(g)} \rightarrow 2Ag_2O + 2SO_{2(g)}$$

To isolate the silver, carbon in the form of charcoal is added.

Nothing is ever as easy as it looks, however. Ores often contain a mixture of metals—as silver and lead are often found together, the smelting process might not give you a pure element, in which case an additional step is needed. One of the earliest methods of isolating pure silver from a mixture of other metals involved the process of cupellation, named after the container in which it was performed—a cupel, which today we would call a crucible, as it has a similar shape. The process of cupellation is elegantly simple: a mixture is heated at very high temperatures and then air is blown over it. The more active elements oxidize first when they come into contact with the oxygen, forming compounds and thus leaving behind the less active elements. If silver and lead are mixed together and heated strongly in the presence of oxygen, lead oxide forms, leaving behind the noble metal silver. The lead oxide would either vaporize or be absorbed into the porous material of the cupel.

Silver was of great use to the ancients as it could be shaped with ease. It is the second most malleable and ductile element, with gold being the first. Although silver is a very soft metal, it is still a little harder than gold; silver's hardness ranges anywhere from 2.5–4 on Moh's sale (compared to gold's 2.5–3). As its melting point (962 °C) is about a hundred degrees lower than that of gold, silver is easier to liquefy. The density of silver (10.5 g mL^{-1}) is almost half that of gold, taking less effort to transport and thus historically making it a more popular choice for routine transactions.

Silver has a wide variety of uses. Owing to its great reflectivity, the best mirrors were once made from silver. We still refer to our metal eating utensils as silverware even though most of our everyday utensils don't contain any silver. Silver is the best electrical conductor, finding its home in a wide array of electronic devices. Touchscreen compatible gloves that enable you to operate your smartphone in cold weather are often made with silver coated nylon fibers.

When photographic film was in its heyday it provided the single greatest use of silver. Silver halide crystals, being light sensitive, are incorporated into the emulsions that coat photographic films and papers. Owing to their light sensitivity, solutions of silver compounds must be stored in opaque bottles to prevent their breakdown by light. If working in the lab you can always tell when you have spilled some silver nitrate on your skin—as soon as you walk outside it turns black. The high intensity UV radiation present in sunlight hastens the reduction of Ag^{+} ions in the silver nitrate into elemental silver. The black stains on your skin are nanosized particles of silver, which fortunately wear off after a day or so.

Perhaps the strangest use of silver is the practice by some of ingesting small amounts of nanosized silver particles in a colloidal suspension, often on a daily basis. Colloidal silver has been touted as an antibiotic, as well as a natural way to boost the immune system. Its proponents claim it can cure everything from diabetes to emphysema. It has even been touted as a treatment for serious conditions such as HIV and cancer. However, there are no scientifically established health benefits from consuming silver, as silver plays no biological role in the body at all. Repeated ingestion can cause argyria, a permanent bluish gray discoloration of the skin, as well as possible kidney damage. The term "blue blood" arose from the aristocracy sometimes having a bluish cast to their skin from having ingested silver particles, as only the wealthy could afford to eat and drink from silver utensils.

Silver does have some legitimate medical uses. The antimicrobial properties of silver have long been known. Since ancient times drinking out of silver vessels has been seen as a sacred rite which could protect the user from illness. The digging of a new well would often be celebrated by throwing in a few silver coins to keep the water fresh. Dropping a silver coin in a bottle of milk would keep it from spoiling. Pioneers would add silver to barrels of water to keep them fresh. Today, silver salts are sometimes used in swimming pools instead of chlorine. A caustic pencil contains a mixture of solid silver nitrate and potassium nitrate—when applied to warts or other skin growths it kills unwanted tissue when it comes into contact with moisture. Even some acne medications utilize colloidal silver.

For many years it was common practice to put a drop or two of silver nitrate solution in the eyes of newborns to prevent bacterial infection, a practice that has been largely abandoned in favor of safer, less caustic medications. If you read the label on your deodorant you might find that it contains silver, either in colloidal form or as a compound such as silver chloride to kill bacteria. Silver Shield, Cool Silver and Silver Protect are a few brands that contain silver. Nanoparticles of silver have been used in a variety of products, including clothing, medical masks, toothpaste, shampoo, towels, and even teddy bears.[8] Even though it is silver ions (Ag^+) and not elemental silver that possess antimicrobial properties, the large surface area of nanosized particles lend themselves readily to oxidation into ions.

Some types of bacteria are, however, resistant to the effects of silver. A silver resistant strain of *Pseudomonas stutzeri* has been found in a silver mine in Canada. It is immune to silver's antimicrobial effects. Pure silver crystals of up to 200 nm in size have been found within the cells of these bacteria, nestled within the periplasm between the cell wall and plasma membrane, cordoned off from the rest of the cell to prevent them from doing harm.[9]

Perhaps one of the more intriguing uses of silver is for cloud seeding. Before the opening ceremonies of the 2008 Beijing Olympics, over a thousand rockets were fired into the clouds, each dispersing silver iodide. The clouds poured forth rain, ensuring that the opening ceremonies would be precipitation free. Although the science behind this method is sound, it is never possible to say with certainty that cloud seeding makes it rain. You cannot conduct a controlled experiment with cloud seeding on a large scale because no two clouds are ever alike—there are simply too many variables to consider; if no two snowflakes are alike then certainly no two clouds are. If I seed a cloud and it rains I can never say with certainty that the seeding caused the precipitation—it might have rained anyway.

In order for precipitation to form there must be something for water vapor to latch onto—a nucleation site. Water vapor can condense, or deposit, on any of a variety of tiny particles as long as they are of sufficient size, typically less than a micrometer. Silver iodide works so well because its crystal structure mimics that of ice. The hexagonal crystalline structure of silver iodide is a dead ringer for an ice crystal, making it very effective at seeding clouds, at least theoretically so.

The first person to come up with the idea of using silver iodide to make it rain was chemist Bernard Vonnegut, older brother to the much more famous Kurt Vonnegut, the novelist. Evidently Bernard's ideas served as the inspiration for the infamous ice-nine substance in Kurt's novel *Cat's Cradle*. In the novel, ice-nine was a special type of ice that melted at 46 °C instead of 0 °C, acting as a nucleation site for any body of water with which it came into contact, instantly freezing it, with potentially disastrous consequences for the entire planet. In the novel, ice-nine was originally developed as a way for soldiers to dry up any mud they happened to be slogging through.

1.8 COPPER: THE HUMBLEST OF THE NOBLE METALS

Even though wars have been fought over silver and gold, copper is the real workhorse of the coinage metals. Copper's symbol, Cu, is based on *cuprum*, the Latin word for copper. This word is a derivation of the Latin word *cyprium*, which is named after the island of Cyprus, which was the main source of copper during the Roman Empire.

Although one can never be certain, copper may have been the first metal discovered and used by man. Copper is more prevalent in the earth's crust than gold and silver combined. Of the 78 elements found in the earth's crust, copper is the 26th most abundant (silver is 65th while gold is 72nd; the rarest is osmium). The early Mesopotamians were skilled at mining and working copper, perhaps as early as 4000 BC. Like silver and gold, copper can be found as a native mineral, but more often than not it must be extracted from its ores. Copper ores are often blue (azurite), green (malachite), or bluish-green (turquoise).

Separating copper from its compounds can be surprisingly easy. If aluminum foil is added to a solution of copper(II) chloride, it will immediately react, as described by the following equation:

$$3CuCl_{2(aq)} + 2Al_{(s)} \rightarrow 2AlCl_{3(aq)} + 3Cu_{(s)}$$

An impressive way to witness this reaction is to sprinkle some solid copper(II) chloride in an aluminum pie pan. If a little water is added a violent exothermic reaction occurs—so much heat is produced that the water instantly boils. The metallic smell of copper is immediately evident as it precipitates out of the solution. As aluminum is a more active metal than copper (aluminum is more easily oxidized than copper), it takes the place of copper and bonds with the chlorine, leaving behind elemental copper. The copper formed is bright red, but quickly turns black as it oxidizes upon exposure to oxygen.

This same reaction can be used to reveal the plastic liner on the inside of an aluminum can. If you sand off the paint on an aluminum can and then immerse it in an aqueous solution of copper(II) chloride, the aluminum will react with the solution, leaving behind the plastic lining. This inert liner is used to prevent the acidic contents of the beverage from reacting with the aluminum of the can.

If you don't have access to copper(II) chloride, equally impressive results can be achieved using copper(II) sulfate, which can be purchased in any hardware store as a root killer. If aluminum foil is added to an aqueous solution of copper(II) sulfate nothing happens, but adding a little sodium chloride will immediately cause a reaction to occur, which produces copious amounts of pure copper.

Copper is the only metal besides gold with a discernible color. As we have seen, copper's ground state electron configuration contains a filled d subshell, which is very stable. However, when light strikes the metal some of its d electrons get excited and jump up to a higher energy level. The energy that these electrons must absorb to make the jump from a 3d orbital to a 4s orbital corresponds to that of orange light. If orange light is subtracted from white light you end up with copper-colored light, thus explaining why copper is copper-colored.

Like gold and silver, copper is soft and workable. On Moh's scale of hardness, copper comes in at around 3, tending to be a little harder than either silver or gold, making it more durable.

Copper is also an excellent conductor, which combined with its ductility makes it well-suited for use in a wide array of electrical and mechanical devices. The average automobile contains over 20 kg of copper, with over a kilometer of copper wiring! Copper has supplanted lead as the material of choice for plumbing, owing to its corrosion resistance. Pots and pans are often made from copper, its aesthetic appeal making it a chic option for a number of other kitchen accessories. The slang term *cop* that is used for police officers may have arisen because of the copper buttons once used on their uniforms, with *cop* being a shortened version of copper.

1.9   COPPER KILLS GERMS TOO

Like silver, the antimicrobial effects of copper have been known since ancient times. It was noted long ago that water transported in copper vessels would stay fresher longer than those transported in other types of containers. As far back as the 17th century BC copper has been used to keep drinking water fresh, receiving a mention in The Edwin Smith Papyrus, the world's oldest surviving treatise on surgery. This ancient treatise recommends using copper as a way to disinfect open wounds, suggesting that copper-alloy swords be filed off over a chest wound so that the shavings would fall in and sterilize the wound. Hippocrates, the father of medicine, wrote of using copper to treat ulcers. His fellow Greeks would sprinkle powders of copper oxide and copper sulfate into open wounds. The green eye shadow favored by Cleopatra did more than convey an appearance of royalty—the malachite $(Cu_2CO_3(OH)_2)$ that imparted the intense green color had antimicrobial properties as well. The Egyptians applied malachite to the skin of burn victims to stave off infections, long before germ theory was recognized.

During the cholera epidemics that raged in France during the 19th century, workers in the copper mines were usually immune to the disease. A similar observation of Finnish copper miners in the 1940s showed that they had markedly low levels of rheumatoid arthritis compared to the general population.[10] These observations and others have led to the lucrative business of selling copper bracelets for arthritis relief. Millions of these bracelets have been sold, despite numerous controlled studies showing no correlation between the wearing of these bracelets and the relief of arthritis symptoms.

Although copper's ability to cure some ailments has been overstated, it is a necessary mineral in the human body (unlike gold and silver), but only in trace amounts. The recommended daily allowance for copper is only 0.9 mg per day, a minuscule amount. We typically get that and more from the foods we eat, with organ meats, seafood, potatoes and nuts being the best sources of copper. Copper is a crucial component of several key enzymes in the body, helping to regulate our metabolism, produce hormones, fight infections and repair damaged tissue. Copper aids in the production of red blood cells, iron absorption, and the proper functioning of the heart, brain and kidneys. The copper concentration in our blood is a respectable 1 ppm.

Some animals have a much higher concentration of copper in their blood than ours, namely those with blue blood. Octopuses, squids, lobsters, horseshoe crabs, and spiders all bleed blue. Whereas the primary oxygen binding protein in our blood is hemoglobin, with an iron atom placed squarely in its center, these bluebloods instead contain hemocyanin, with two copper atoms per molecule. Hemocyanin is not bound to the blood cells but rather floats around freely, and when deoxygenated is colorless.

As essential as copper is to life, too much can be deadly. Rat LD_{50} values (lethal dosage required to kill 50% of a test population) for oral consumption are notoriously low. Copper(II) chloride has an LD_{50} of 584 $mg\,kg^{-1}$, but for copper(II) sulfate it is only 300 $mg\,kg^{-1}$.

The toxicity of copper has been exploited by homeowners and gardeners. Copper sulfate crystals are the material of choice to kill overgrown roots in sewer lines. Bordeaux mixture, which originated in the Bordeaux region of France in the 19th century, consists of copper(II) sulfate and lime $(Ca(OH)_2)$ and is still used as an agricultural fungicide.

Julia Child was one of America's best loved chefs, enthusiastically bringing French cuisine into countless homes *via* her cookbooks and television programs. However, long before she became a famous chef Child worked for the Office of Strategic Services during World War II, an organization that would eventually become the CIA. Child was part of a team tasked with developing an effective shark repellant for the Navy, which was searching for ways to protect downed servicemen in the ocean. Although copper sulfate showed a lot of promise, they eventually settled upon copper acetate as the active ingredient in their shark repellant. A recent news report released by the CIA described the repellant:

> To create the repellent, copper acetate was mixed with black dye, which was then formed into a little disk-shaped "cake" that smelled like a dead shark when released into the water. These cakes could be stored in small 3-inch boxes with metal screens that allowed the repellent to be spread either manually or automatically when submerged in water. The box could be attached to a life jacket or belt, or strapped to a person's leg or arm, and was said to keep sharks away for 6 to 7 hours.[11]

The repellant had a 60% success rate, and was used well into the 1970s.

Even pure copper is no slouch when it comes to killing germs, and its use as a disinfectant has been vindicated by modern science. A variety of pathogenic organisms, including bacteria, viruses and fungi, will be dead within two hours if they happen to land on a copper or copper alloy surface. Copper has been registered with the US Environmental Protection Agency as the first solid antimicrobial substance. Many nasty bugs such as *E. coli, Staphylococcus, Pseudomonas* and a whole host of others die within an hour or so after landing on a copper surface. One study showed that after 24 hours 500 000 viral particles of influenza A remained on a stainless-steel surface, but after six hours on a copper surface only 500 viral particles survived.[12] One recent study showed that each year in the UK 300 000 people pick up infections from hospitals, of which 5000 die annually.[13] Hospitals and health care facilities have recognized the value of copper's antimicrobial properties, using copper alloys on doorknobs, bed rails, IV poles and anywhere else where pathogens can gain a foothold.

Copper's toxicity to pathogens is due to its oxidative ability. As copper is readily oxidized, it easily gives up its electrons. If copper is oxidized by a reducing agent, electrons are removed from the copper. If bacterial cells, for example, encounter a copper surface, their bacterial membrane is compromised as copper ions pour through holes in the cell membrane. These copper ions act as a catalyst, converting hydrogen peroxide (a waste product of cellular metabolism already present in the cell) into the highly reactive hydroxy radical ($HO^\bullet$), which kills the cells.[14]

The reactivity of copper ions makes it effective for a variety of biocidal purposes. Ship hulls are subject to a wide variety of biological attacks from marine life that can accelerate the rate of corrosion of the metal. Copper is often incorporated into the paint that is used on these hulls, which is often red owing to the presence of copper(i) oxide. One target of this antifouling paint is the barnacle, which can wreak havoc on a ship's hull. These little crustaceans spend the first part of their lives looking for a surface to adhere to, and once they find it they attach themselves for life, using a natural superglue that is extremely tough. (This glue is so strong and so effective underwater that it has been studied extensively in an attempt to replicate it synthetically.) Antifouling paint is not cheap, however, and can cost upwards of $100 USD per liter.

As Freud observed, money is widely considered to be filthy, hence the term "filthy lucre". There is some truth to this assessment, as money exchanges hands quite often, and you really can't know everywhere it has been. However, the idea that money is hopelessly germ infested may best apply to paper money and not coinage. If anything, coins probably have fewer germs on their surfaces than almost anything else lying around, especially if the coins are made from silver or copper. Actually, most metals possess some antimicrobial properties—the scientific term for this property is the oligodynamic effect.

1.10   ALLOYS: MAKING METALS EVEN BETTER

During the Bronze Age man began alloying metals together, dramatically increasing their usefulness. In their elemental form, pure metals are surprisingly limited in their use. Any metal in use today will almost certainly be an alloy. An alloy is a homogeneous mixture of two or more elements, with at least one of them being a metal. By mixing metals together, they can be made harder, stronger and more corrosion resistant.

Gold, for example, is very soft by itself. The established unit for determining the purity of gold is the karat. Gold jewelry typically comes as either 14-karat or 18-karat, the karat being a measure of how much gold is in the sample. Pure gold is 24-karat; the percentage of gold is easily calculated by dividing the karat amount by 24, so the percentage of gold in a 14-karat piece is found by dividing 14 by 24, which is 58%. The remainder is mostly silver, with perhaps some copper thrown in for good measure. Typically, the legal minimum allowable amount of gold in jewelry is 10-karat, but in some countries, such as Greece and Denmark, 8-karat jewelry has at times been permitted. Even though 24-karat gold is so soft that it can easily be bent, there is still a market for it, especially in China, where pure gold is viewed as a status symbol. It can also double as currency, as it is just as pure as any other sample of gold.

Even the gold medals awarded during the Olympics are not pure gold, but rather are mixtures, with just a little gold on the surface. The medal itself has a mass of 500 g, but contains just

6 g of gold, in the form of a very thin plating which makes up just 1% of its composition. The value of one of these gold medals is $565 USD; if made from pure gold it would be worth over $20 000! The last time an Olympic gold medal was mostly gold was in 1912, when the Olympics were held in Stockholm.

The Olympic silver medals, though, are almost pure silver, and are worth a little over $300 USD. The "bronze" medals are a true alloy, with about 95% copper and 5% zinc; this composition actually makes the medals brass, as bronze contains tin as well as copper and zinc. (Numismatists differ from chemists in their definition of brass, typically requiring at least 10% zinc to qualify as brass; any coins with less than 10% zinc are classified as merely copper.) These metals have a net value of just over $2.00 USD. The composition of the medals changes from one Olympics to the next, with the host city's organizing committee having the final say in the exact composition of the medals, as long as certain standards set by the International Olympic Committee are met. Gold medals have not always been awarded to the winners. In the first modern Olympics, held in 1896, the winners were awarded silver medals, with the runners-up receiving bronze.

1.11 THE TWO TYPES OF ALLOYS

Alloys are typically of two types—either substitutional or interstitial. A substitutional alloy forms when one metal is replaced with another that is similar. Brass and bronze are good examples, in that zinc or tin atoms take the place of the copper atoms. Metallic bonding is at work in these types of alloys, in which neighboring metal atoms share electrons freely in the vast sea of delocalized electrons that surround their nuclei. This electron-free-for-all is responsible for many metallic characteristics that alloys share with pure metals, such as malleability and conductivity.

Pure metals contain layers of metal atoms arranged in regularly repeating crystalline patterns that are arranged in stacks, easily sliding past one another, making them soft and easily deformed. However, in an alloy, this regular repeating pattern is interrupted by the addition of different metals of various sizes. The smooth layers present in pure metals are disrupted by irregularities in the alloy's surface, making it more difficult for the layers to slip past one other. Even though the resulting structure tends to be harder, the bonds between neighboring atoms are actually weaker, as there is no substitute for the real thing in making a structure sturdier. Imagine taking a brick wall and substituting some of the bricks for other bricks of different sizes. The resulting wall might be stronger in some ways, but weaker in others.

Alloys are harder than pure metals but will always have a lower melting point than either of their constituent elements. A 50/50 solder, for example, is made from an alloy of 50% lead and 50% tin—it's very low melting point makes it an excellent choice for joining together sensitive electronic components. Wood's metal, an alloy of varying composition containing bismuth, lead, tin and cadmium, has a very low melting point and is used in fire sprinklers, melting when the temperature becomes high enough to activate it. The composition can be varied depending on where the sprinkler is placed—a higher activation temperature would be needed in a kitchen for example, in which high temperatures are the norm, as opposed to, say, a classroom. The melting point of Wood's metal can be brought down to as low at 70 °C, far lower than the melting point of any of its constituent elements.

The other type of alloy is interstitial, which occurs when very small atoms wedge into the spaces between the atoms in a metallic crystalline lattice. Think of them as the mortar between the bricks. These smaller atoms tend to be nonmetals. The nonmetals, being on the far right of the periodic table and thus of diminutive size, are most likely to be the atoms wedged into the spaces between the metals in an alloy (Figure 1.5).

Steel is a common example of an interstitial alloy, in which carbon or sulfur atoms are added to iron, making a much harder product. However, if too much is used the hardness changes into brittleness, which is bad. The trick is to add just enough nonmetal so as to achieve the right mix of hardness and malleability. Samples of steel recovered from the wreck of the Titanic, which sank in 1912, revealed that the steel used to make it was ten times more brittle than steel used today. However, this type of steel was acceptable at the time the Titanic was built. Tests revealed it to be high in sulfur, contributing to the brittleness, and low in manganese (manganese tends to make steel more

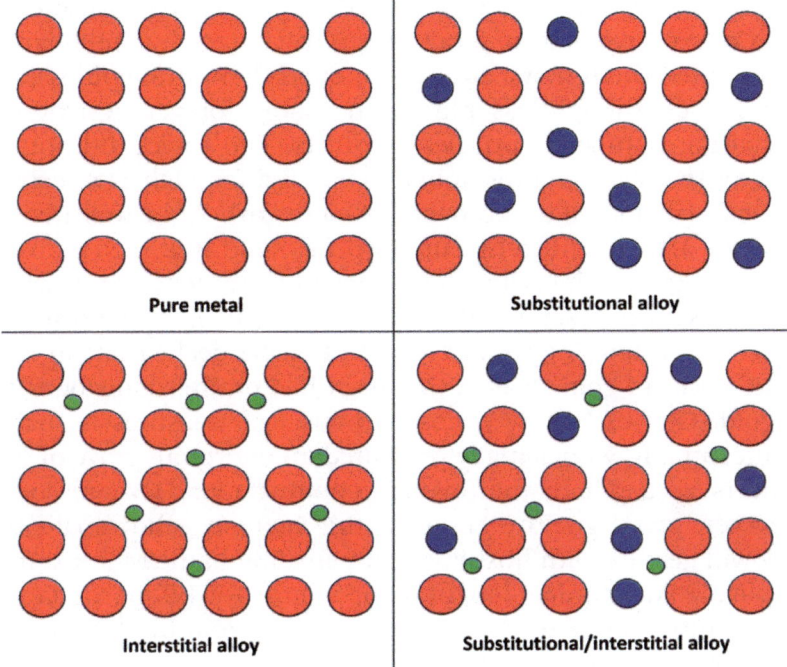

Figure 1.5 Types of alloys.
Image courtesy of Wikipedia/CC0 1.0.

malleable). Examination of the rivets, which were used to hold the hull together, contained abnormally high levels of slag, the glassy residue left over from the refinement of iron ore, making them quite brittle as well.[15] Despite its veneer of invincibility, the Titanic was essentially an accident waiting to happen.

The first alloys used by ancient societies were naturally occurring. It is likely that the first metals ever found by man were meteorites found lying on the surface of the ground. These metals from space are usually iron-nickel alloys. Among the oldest objects fashioned from meteorites are beads recovered from a 5000-year old Egyptian tomb.[16] Trace amounts of germanium, cobalt and phosphorous in the beads strongly indicate an extraterrestrial origin, indicating that they were likely fashioned from an iron-nickel meteorite. Many other ancient artifacts such as tools and weapons were made from meteorites as well, long before the advent of the Iron Age. They were often used in religious and tribal ceremonies, their fall from heaven testifying to their divine origin.

Early Americans were also very adept at working with metals. Tons of gold relics were looted from the Aztecs by the Spanish, who melted them down and refashioned them into gold bars or coins. Such thoughtless plundering sadly forever destroyed a staggering number of priceless artifacts. The Aztecs did not use gold for money—gold was much too sacred for such a pedestrian purpose (that's what cacao beans were for). Gold was instead used for artistic and religious purposes, and was seen as a gift from the gods—the Aztec word for gold was teocuitlatl, which translates into "excrement of the gods" (furthering bolstering Freud's claims). The royalty would often be decked out in elaborate pieces of gold jewelry; they even had entire rooms outfitted in gold. One of history's most famous lost treasures is that of Montezuma II, who actually presented an enormous trove of gold artifacts as a gift to the Spanish conquistador Hernan Cortes. However, greed got the better of the Spanish, who repaid this goodwill gesture with hostility; in an ensuing battle Montezuma's massive gold hoard—thought to be worth 3–4 billion dollars in today's money, was recaptured by the Aztecs and hidden—to this day it has not been found.

These early Americans were such skilled metallurgists that what the Spaniards thought was real gold was often an alloy of gold. When they discovered the truth, they coined the term tumbaga, which is based on the Malay word *tembaga*, meaning copper or brass. Tumbaga is not a technical term, but rather can refer to any alloy of gold or silver mixed with copper—percentages of each can vary widely. Although some of the artifacts plundered from the Aztecs were pure gold, many others were not. Archaeological evidence has revealed advanced copper smelting operations as early as 900 AD in the Mexican state of Michoacán, revealing a sophisticated level of metallurgical expertise. It is much more difficult to work with copper than gold, as gold is simply mined from the earth in its native form while copper is usually extracted from its ore.

Many of these gold-colored tumbaga artifacts were easily mistaken for pure gold because their true identity was skillfully concealed by a layer of pure gold on the surface. Gold was not added, however, but rather other metals were removed in a process known as depletion gilding, a skill at which the pre-Colombian Americans excelled. The alloy would be immersed in an acid or other compound that would react with copper but not gold. The oxidized copper would then be abraded off, leaving behind pure gold. This process would be repeated numerous times, making the gold layer successively thicker. One method of depletion gilding involved rubbing the tumbaga alloy with the sap from the *Oxalis* plant, which as its name implies contains a high concentration of oxalic acid ($C_2H_2O_4$). Once it was thoroughly coated the metal would be heated. This process would be repeated numerous times until the acid was concentrated enough to react with the copper.[17]

Although Montezuma's famous treasure trove is still out there waiting to be discovered, relics from his era are still being found, often in their recast form. In 1992, a horde of silver bars were recovered from a 16th century Spanish shipwreck off the coast of the Bahamas. In all, over 200 crudely formed cast slabs were discovered that were likely made from melted down Aztec artifacts. Small portions of the bars that were broken off and analyzed revealed their composition to indeed be tumbaga—they were made from 50% silver and 50% copper.

1.12 WHY WERE COINS NEEDED?

No government needed to declare silver and gold valuable—they were already the de facto currency for much of the civilized world before any entity officially declared them so—it's as if their allure had been somehow hardwired within the brain. It was only a matter of time before a ruling class would capitalize on this weakness for shiny objects and use it to wield its power. The first kingdom to do just that was Lydia, in the 7th century BC. Today Lydia is part of Turkey, but at one time it was a separate empire. Their rise to greatness was partly a result of their creation of the modern concept of money, a concept that changed the world forever, namely that governments determine the medium of exchange—the governing authorities set the values and make the rules. In order to do such a thing standardization needed to occur. As it was not practical for every citizen to carry around a scale to weigh their metals, if these metals could be fashioned into uniformly recognizable sizes and shapes then they could be exchanged freely without the use of a scale. So, we witness the birth of coins way back in this ancient kingdom, coins that could not be made by just anyone—they were to be issued only by the government and it was only by governmental decree that they were deemed official currency.

The development of coinage greatly streamlined the process of buying and selling. Before the advent of coins, paying for items using precious metals was a very time-consuming process, involving a lot more rigmarole than just weighing the metals. The first step was to determine the purity of the sample being presented. The Lydians developed an ingenious method to assay the various metals which were offered as payment. The purity of a metal was determined by using a touchstone—a shiny black rock similar in composition to black jasper (a type of quartz). The sample would first be rubbed on the touchstone, and the color of the mark left behind would be compared to the color of the mark made by a sample of known composition. A Lydian merchant would typically have a set of 24 samples, each consisting of a different combination of metals mixed with gold—each one would be rubbed on the touchstone and then compared with the mark left by the unknown metal, enabling him to determine its precise composition.

Once the purity of a precious metal was established, the next step was to determine its exact mass. Unscrupulous dealers would attempt to cheat their customers by using inaccurate weights on their scales, hence the strong admonitions in the Bible warning of such a practice. Proverbs 11:1 states:

> A false balance is an abomination to the Lord, but a just weight is his delight.

It was common practice to carry around different sets of weights, one accurate and one not so much. The weights you decided to use would depend on the age and experience of your customer. Hence, this additional admonition in Proverbs 20:10:

> Differing weights and differing measures, both of them are an abomination to the Lord.

The exact date of the first coin ever minted is lost to history, but scholars generally agree that it was made from electrum, a naturally occurring alloy. Electrum is composed of gold and silver, with the composition varying depending on the source, but it was usually close to 50 : 50. It was often called white gold, as lower concentrations of gold made it lighter.

Electrum derives its name from the Greek word *elektron*, which is the same word used for amber, the yellow fossilized tree sap that readily attract electrons, making it an excellent producer of static charge. Electrum derived its name from amber not because it shared any of its properties, but only because it had the same color.

The Lydians obtained the bulk of their electrum from the Pactolus River. Often a sheepskin would be used to sift the sand and gravel, in the hopes that some gold would be left within its fibers. This ancient method of filtration is likely the origin of the story of Jason and the Golden Fleece. The gold that was found in abundance was believed to have been from King Midas, who washed off his golden curse by bathing in this river. A more scientific explanation has the gold alloy being carried downriver from the hillsides of nearby Mount Timolus, which contained substantial deposits of electrum.

1.13 ARCHIMEDES AND THE GOLDEN CROWN

This ability of gold and silver to alloy together so easily has provided the stuff of many legends. Perhaps the most famous is the story of Archimedes and the golden crown. Archimedes was a Greek mathematician and scientist who lived in the 3rd century BC. King Hiero II of Syracuse had a crown made that was supposed to be pure gold. However, he did not trust the goldsmith who had made it, and became convinced that he had somehow been cheated. He suspected that the crown may have had some less expensive silver mixed in, but there was no way he could verify his hunch. He became obsessed with trying to prove that the crown was not pure gold. He eventually enlisted the help of Archimedes, who gladly accepted the assignment. The one condition was that he could not harm the crown in any way. Archimedes puzzled over how best to solve this problem. He always thought better when soaking in the tub, so he drew a bath. While lowering himself in the tub he noticed that the water level rose, and that's when he got the idea that water displacement could help solve the riddle. He was so excited at his insight that he immediately ran naked through the streets yelling, "Eureka!" which translates to "I have found it!".

So, Archimedes began his little experiment with the crown by first taking its weight, and then recording how much water it displaced. As we saw with the cacao bean, water displacement can be an effective method in helping to determine the density of an object. If the crown were pure gold, every 19.3 g of it would displace exactly 1 mL of water. If the crown weighed 1000 g it would displace 51.8 mL of water (1000/19.3). If the crown were made of pure silver, with a density of 10.5 $g\,mL^{-1}$, then a 1000 g crown would displace 95.2 mL of water (1000/10.5). Therefore, if a 1000 g crown displaced anything more than 51.8 mL of water, then it was probably not pure gold. According to legend, the crown did indeed displace more water than that of an equal amount of gold. The king was happy, the goldsmith was probably beheaded and Archimedes' status was elevated.

The story of Archimedes and the crown is recounted by the architect and military man Vitruvius in the 1st century BC, in Book IX of his ten-book series on architecture. Scholars who have closely examined this account claim that the type of lab equipment available to Archimedes would not have enabled him to detect such a small difference in water displacement between the crown and a sample of pure gold, as the goldsmith probably would have been careful not to make his fraud too obvious so as to preserve his head. It is highly likely that the passage of time may have corrupted some details of this story, but it has remained in popular lore because theoretically it is possible—diluting gold with silver was a common practice of the day, precisely because gold and silver form such a seamless alloy.

1.14 THE FIRST COINS

Historians believe the first coin ever made by the Lydians had an image of a lion on its obverse side, with the reverse side being blank (Figure 1.6). In numismatic terms, what we commonly refer to as "heads" is the obverse side of a coin, and "tails" is technically known as the reverse side. The lion was the official symbol of the ruling dynasty at the time—even though lions today are mostly confined to Africa, they once roamed throughout southern Europe and the Middle East all the way to India.

Figure 1.6   Lydian coin.
Image courtesy of the Classical Numismatic Group, LLC/www. cngcoins.com.

These first coins were not quite round—they were actually more of an oval shape, with a mass of just under 5 g and a diameter of 13 mm at their widest point. They were made of electrum, with a composition of 55% gold, 43% silver and 2% copper. Today, one of these coins would cost around $2000 USD; at the time it was made it would have been worth a month's wages.

These first Lydian coins would hardly fit the definition of a modern coin. We expect our coins to be flat and uniform. But many of the Lydian coins were rather like lumps of metal, resembling more a misshapen piece of pottery than official currency. One of these first coins bears a striking resemblance to a human ear. As these earliest coins were all hand-made each was unique—a quaint feature of all ancient coins.

The first step in making these coins was acquiring the raw material, usually the naturally occurring electrum. Just the right amount of metal had to be fashioned together to make a "blank" coin before it was imprinted. These first blanks were made by melting down electrum and then pouring the molten metal into a mold. This process is known as casting, which has a long and rich history. Casting has been used for thousands of years to make sculptures, tools and weapons. It is still widely used today to make jewelry, auto parts, plumbing fixtures and a plethora of other products. If you have an object made of metal, chances are it is a product of the casting process. Casting is not hard to do— if you have a mold, a blowtorch and some metal you can melt it, pour it into a mold and make a cast. When it cools and hardens you pop it out of its mold.

The relatively low melting points of gold and silver make them perfectly suited for making coins—pure gold melts at 1063 °C and pure silver at 961 °C. There is a captivating beauty in seeing molten metal. It appears especially lustrous as the surface oxidation is removed, revealing a shimmering beauty underneath. Molten metal looks very much like mercury. A typical propane torch can attain temperatures close to 2000 °C, which is more than adequate to melt gold or silver. A 50/50 mixture of gold and silver will melt at around 860 °C. A 70/30 mixture of gold and silver will have the lowest melting point, at around 800 °C. Like all alloys, a mixture of metals will always have a lower melting point than either of its constituent elements.

In coin-making, however, only the blanks were made by the casting process—not the final coin itself. If casting is used to make a complete coin, there is a good chance it is counterfeit—counterfeiters have long relied on the casting process to make illegal replicas of official coins.

Once a blank is made it needs to be struck, which requires the making of a die. Ancient dies were simply pieces of cylindrical metal with a wide end and a narrow end, with a design engraved on the wider end. These dies were usually made from either bronze or iron. Bronze is softer than iron and thus easier to engrave, but the bronze, being softer, tends to wear out faster.

After the die is engraved, the next step is to impress the design onto the coin blank itself. The die would be mounted into an anvil and then the blank placed on top of the die. Another piece of cylindrical metal with a flattened end, known as a punch, was then placed on top of the blank coin and struck with a sledgehammer, engraving the image onto the coin. These earliest coins were only struck on one side; eventually coins would be struck on both sides, with the upper punch also serving as a die, which had a separate image engraved onto its end.

While this method seems very labor intensive, and indeed it was, it was still quite efficient. One coin-maker who wanted to replicate these ancient methods discovered that he could strike 100 coins per hour using this technique, and a recent research study suggested that a team of ancient coin-makers could strike 20 000 coins in one day.[18]

The Lydians did not just make one type of coin—they varied the weights to make coins of different values. They named their coins *staters*, with a one-stater coin weighing 14.1 g. Various denominations of lesser value were also made, such as half-staters, third-staters, and so forth, all the way down to 1/96 staters, which only weighed 0.15 g. This uniformity of weight was an important feature, perhaps the most important, in helping coins gain widespread acceptance.

Scholars have long debated whether or not the Lydians used naturally occurring electrum, or whether they added additional silver to make the coins cheaper to produce. Regardless, a common complaint about these coins was that they contained too much silver and not enough gold. To counter this problem, when King Croesus, the last king of Lydia, came to power in 560 BC he confiscated all existing coins and had them melted down, reissuing all coinage as either pure gold or pure silver. This act created the first bimetallic coinage system, with gold being worth 10 times as much as silver. The new coins were stamped on both sides, with an animal on one side and the value incused on the reverse side. A coin is incused when the impression is depressed or concave, as opposed to being raised or convex. Most coins have a raised image—depressed images are rare.

A wealth of archeological evidence points to Lydia as indeed being the birthplace of coinage. Hundreds of different types of Lydian coins have been discovered. The 1905 excavation of the Temple of Artemis in ancient Ephesus yielded 93 electrum coins, all estimated to be from around 630 BC. These coins were a diverse lot—some were mere lumps of electrum while others were stamped with various animal images, including roosters, beetles, horses and goats. The wide variety of shapes and sizes indicates that standardization had not yet occurred. Yet the coins themselves are remarkably well preserved, resisting the ravages of corrosive atmospheric oxidation for 2600 years, providing ample evidence for gold and silver's nobility.

REFERENCES

1. S. Ferranti, With Cigarettes Banned In Most Prisons, Gangs Shift From Drugs To Smokes, *Daily Beast*, 2003, https://www.thedailybeast.com/with-cigarettes-banned-in-most-prisons-gangs-shift-from-drugs-to-smokes.

2. Safety assessment for chocolate. BC Centre for Disease Control, Provincial Health Services Authority. Environmental Health Services. Food Issue – Notes from the Field. http://www.bccdc.ca/resource-gallery/Documents/Educational%20Materials/EH/FPS/Food/fmchocolate1.pdf.

3. K. Aleklett, D. J. Morrissey, W. Loveland, P. L. McGaughey and G. T. Seaborg, Energy dependence of ^{209}Bi fragmentation in relativistic nuclear collisions, *Phys. Rev. C*, 1981, **23**, 1044.

4. Z. Liu, M. Frasconi, J. Lei, Z. J. Brown, Z. Zhu, D. Cao, J. Iehl and S. J. Fraser, Selective isolation of gold facilitated by second-sphere coordination with α-cyclodextrin, *Nat. Commun.*, 2013, **4**, 1.

5. L. Bütof, N. Wiesemann, M. Herzberg, M. Altzschner, A. Holleitner, F. Reith and D. H. Nies, Synergistic gold-copper detoxification at the core of gold biomineralisation in Cupriavidus metallidurans, *Metallomics*, 2018, **10**(2), 278–286.

6. H. Knosp, R. Holliday and C. Corti, Gold in dentistry: Alloys, uses and performance, *Gold Bull.*, 2003, **36**(3), 93–102.

7. D. S. Girling, Gold plating in submarine telephone cable repeaters, *Gold Bull.*, 1973, **6**(3), 69–71.

8. T. Benn, B. Cavanagh, K. Hristovski, J. D. Posner and P. Westerhoff, The Release of Nanosilver from Consumer Products Used in the Home, *J. Environ. Qual.*, 2010, **39**(6), 1875–1882.

9. T. Klaus, R. Joerger, E. Olsson and C.-G. Granqvist, Silver-based crystalline nanoparticles, microbially fabricated, *Proc. Natl. Acad. Sci. U. S. A.*, 1999, **96**(24), 13611–13614.

10. J. R. Sorenson, Copper complexes offer a physiological approach to treatment of chronic diseases, *Prog. Med. Chem.*, 1989, **26**, 437–568.

11. CIA News and Information, Julia Child and the OSS recipe for shark repellant, 2015, https://www.cia.gov/news-information/featured-story-archive/2015-featured-story-archive/shark-repellent.html.

12. J. O. Noyce, H. Michels and C. W. Keevil, Inactivation of influenza A virus on copper versus stainless steel surfaces, *Appl. Environ. Microbiol.*, 2007, **73**(8), 2748–2750.

13. House of Commons Public Accounts Committee, *Reducing Healthcare Associated Infection in Hospitals in England. Fifty-second Report of Session 2008–09*, 2009, p. 3.

14. Z. D. Jovanovic, M. B. Stanojevic and V. B. Nedeljkov, The neurotoxic effects of hydrogen peroxide and copper in Retzius nerve cells of the leech Haemopis sanguisuga, *Biol. open*, 2016, **5**(4), 381–388.
15. J. J. Hooper, T. Foecke, L. Graham and T. P. Weihs, The metallurgical analysis of wrought iron from the RMS Titanic, *Meas. Sci. Technol.*, 2003, **14**(9), 1556–1563.
16. T. Rehren, T. Belgya, A. Jambon, G. Káli, Z. Kasztovszky, Z. Kis, I. Kovács, B. Maróti, M. Martinón-Torres, G. Miniaci and V. C. Pigott, 5,000 years old Egyptian iron beads made from hammered meteoritic iron, *J. Archaeol. Sci.*, 2013, **40**(12), 4785–4792.
17. A. C. Sparavigna, Depletion gilding: An ancient method for surface enrichment of gold alloys, *Mech., Mater. Sci. Eng.*, 2(1), 2016, 99–105.
18. G. Carter and R. Nord, Calculation of the average die lifetimes and the number of anvils for coinage in antiquity, *Am. J. Numismatics (1989–)*, 1992, 3/4, 147–164.

CHAPTER 2

The Rise of Coins

Nothing in the history of humankind, save for the cultivation of noble fire itself, can compare to the discovery of metal. It is a wondrous substance, born of the earth, purified by the air, nurtured in the fire it flows like water. Poured in dazzling brilliance, it takes on the shape of its container. Struck, it yields to the hammer, becoming wand, sword, chalice or shield. Drawn, it becomes as a thread. Polished, it contains the whole world in its reflection. Honed, its edge cuts lesser materials without violence. There are few objects which would not be more durable, more beautiful, and more useful if fashioned by art and ingenuity from metal.[1]

Kevin M. Dunn, Caveman Chemistry.

The earliest coins were a byproduct of the technology of the age. Coins only emerged when metallurgy had advanced to the point at which metals could be molded into useful forms through either casting or forging. Advances in money-making have always paralleled advances in chemistry, advances which are continuing to the present day—indeed, today's innovations will appear on tomorrow's money. The chemical knowledge that the ancients employed had its roots in alchemy, as the earliest chemists were alchemists.

The Chemistry of Money
By Brian Rohrig
© Brian Rohrig 2021
Published by the Royal Society of Chemistry, www.rsc.org

2.1 THE INFLUENCE OF ALCHEMY

Chemistry and alchemy are both based on the root word *Chema*, the name of the book which legend claims represents the ancient secrets of metallurgy that were handed down to men by the angels. Alchemy has always been a hybrid of chemistry and mysticism—early alchemists were obsessed with discovering the philosopher's stone, a magical substance which would not only convert base metals into gold but would also provide enlightenment and even immortality. J. K. Rowling gave a nod to alchemy's influence by naming her first book *Harry Potter and the Philosopher's Stone*, (a title that will be familiar to fans in the UK, whereas the rest of the world read *Harry Potter and the Sorcerer's Stone*).

Alchemy most likely began in Egypt—the earliest alchemists mastered the chemical techniques of mummification in hopes of preparing the deceased for the afterlife. However, it was Greek thought that kept alchemy alive and allowed it to flourish. Ancient Greece produced no shortage of great thinkers, with the ideas of Aristotle, Archimedes, and Pythagoras shaping human thought for millennia to come. The world's first theoretical chemists were Greeks. It was Empedocles in 420 BC who proposed that all matter was made of four elements—earth, air, fire and water. If one considers the four main states of matter—solids, liquids, gases and plasma—this theory holds up remarkably well. However, Empedocles also held other less scientific views, subscribing to reincarnation and postulating that Love and Strife were the two main forces acting in the universe.

Most high school chemistry students are familiar with Democritus, who reasoned that all matter is made up of tiny indivisible particles known as atoms. He speculated that if you were to take any object and continually cut it in half, you would eventually reach a point at which you could cut it no more—this smallest particle of matter he called an atom. Aristotle, on the other hand, argued that matter was continuous, claiming that any object could be divided in half indefinitely. We now know, of course, that the atom is composed of many smaller particles. Scores of different types of subatomic particles have been discovered. It is more than a little ironic that the atomic bomb, which obtains its energy from the splitting of the nucleus, derives its name from the Greek word *atomos*, which means indivisible. Like his counterparts, Democritus held some rather fallacious views. He hypothesized that the properties of macroscopic matter were similar to the properties of the atoms themselves. He thought that solids were composed of hard atoms, oils of slippery atoms, and so forth.

Plato built upon this four-element framework by postulating an additional four qualities—hot, cold, wet, and dry. He claimed that each of the four elements were composed of two of these qualities. Fire was hot and dry and water was cold and wet. Air was hot and wet, but earth was dry and cold. Each element thus shared a property with another element, so if cold and wet water were heated it would turn into hot and wet air, which if dried would become hot and dry fire. If you were to cool this hot and dry fire you would get dry and cold earth, which could then be converted back into water, perpetuating the cycle.

It was Aristotle's propagation of these theories that allowed alchemy to flourish. As it was believed that all substances were composed of varying amounts of fire, earth, water and air, then transforming any substance into another was simply a matter of discovering the precise proportions of each. By finding the correct recipe, you could convert any base metal into gold. It would take 2000 years before this approach would give way to more modern views of the atom, beginning with John Dalton's atomic theory in 1808.

Most traditions held that there was a fifth element as well, referred to as the void or the sky, intimating a spiritual dimension, or perhaps some type of unseen energy force. Aristotle called this fifth element the ether. He was not referring to the anesthetic, but rather to the mysterious invisible substance that permeated the heavens and all space between matter. Medieval alchemists referred to this fifth element as the *quinta essentia*, or quintessence—the search for this substance was the all-consuming passion of the alchemists.

The human soul was made of this fifth element, as were the heavenly realms. However, it also dwelt within all matter. The hard part was isolating it, which involved long hours in the lab where one would be exposed to toxic vapors of lead and mercury, vapors that could drive a man mad. Depending on which alchemist you talked to, this fifth element could be summoned by the philosopher's stone. Others believed that this fifth element *was* the philosopher's stone, and that if found could transmutate any element into gold, not to mention imparting eternal life.

The notion of the ether was so entrenched that it was not until late in the 19th century that its existence was seriously called into doubt. Even Newton entertained notions of the ether, attempting to explain everything from optics to gravity using an ethereal framework. Being an alchemist himself, it should come as no great surprise that Newton would entertain this most alchemical of ideas. However, as he hammered out his theories Newton found he no longer had need for an ether, as forces could act at a distance perfectly well without the use of any conducting medium. Today, we have our own ether—the twin mantras of dark matter and dark energy, the mysterious stuff that allegedly permeates everything, making up an astounding 96% of our universe. It is the gravitational force exerted by dark matter that keeps our galaxies from flying apart, while dark energy is simultaneously driving the expansion of the universe at an ever-increasing rate. No one has ever directly detected dark matter or dark energy (or they would be called something else), yet our theories demand the existence of these twin ghosts in the machine.

The alchemists held that there were seven metals that possessed spiritual significance. Possession of each metal would imbue its owner with special strengths and talents. As the alchemists also dabbled in astrology, they assigned a separate planet to each metal, giving each their own day of the week. Table 2.1 provides a summary of the alchemists' seven most important metals:

Table 2.1 The alchemists' seven most important metals.

Metal	Planet	Day of the Week	Spiritual significance
Gold	Sun	Sunday	Wealth, beauty, healing
Silver	Moon	Monday	Purity, intuition, persistence
Iron	Mars	Tuesday	Strength, courage, aggression
Mercury	Mercury	Wednesday	Death, divination, mystery
Tin	Jupiter	Thursday	Knowledge, wisdom, logic
Copper	Venus	Friday	Creativity, love, compassion
Lead	Saturn	Saturday	Impurity, death, darkness

If you examine the properties of the three coinage metals—gold, silver and copper—you will quickly notice that they are the metals with the most desirable traits. Who wouldn't want the wealth, beauty and healing that gold could impart? It was the next best thing to the philosopher's stone itself. The metals that made up coins were not just utilitarian, their inertness was seen as evidence of a divine essence that dwelt within; gold, being the most inert was thus the closest to perfection. The belief that possessing noble metals would imbue their possessors with certain alluring traits only helped to popularize metallic money.

When Alexander the Great conquered Egypt he created the Library of Alexandria, the greatest library in the world, which was to house all of the world's knowledge, including the alchemical texts. At least part of this library is believed to have been burned, possibly by the Romans, resulting in the tragic loss of many ancient documents. Emperor Diocletian, ever the pragmatist, ordered the burning of all the Egyptian alchemical texts he could find, for fears that if the alchemists did succeed in finding a recipe for gold, then all of Rome's gold would immediately be devalued, destroying the economy.

2.2   THE GREEKS: COINS BECOME ART

Lydia was conquered by Persia during the reign of King Croesus, and later Persia would be conquered by Alexander the Great of Macedonia, but these conquests did not put an end to coinage. Just as the Greeks would assimilate the cultural advancements of their vanquished foes, so too did they do with Lydia's coins. The Greeks improved upon the Lydian methods, striking very beautiful coins that reflected the cultural values they held dear. Although the Lydians may have invented coinage, it was the Greeks that elevated it to an art form. Unlike Lydian coins, Greek coins were always struck on both sides. They were composed of either silver or gold, with silver being more common; copper was also widely used. The drachm was a very common denomination, containing 4.6 g of silver. For larger purchases you could use the tetradrachm, and for really large purposes you could use the decadrachm. With a mass of 46 g, the decadrachm is often described as the most beautiful of ancient coins (Figure 2.1). As Greece was divided into hundreds of relatively independent

Figure 2.1 Silver Syracusan decadrachm depicting winged Victory on quadriga, verso. Commemorative coin minted to celebrate the victory of Gelo and the Syracusan troops over the Carthaginians in the Battle of Himera, 480 BC. Ancient Greek coins, 5th century BC. © DeAgostini/Getty Images.

city-states, each had freedom to produce their own money, leading to a wide variety of circulating coins.

Eventually images of people began appearing on coins, usually that of a Greek god (in the form of a man) or a popular hero. Animals were still commonly used, with the owl being quite popular. Other animals found on ancient Greek coins include eagles, dolphins, horses, bees, bulls, turtles, elephants, scorpions and camels. Plants such as celery, roses and wheat can also be found. The Greeks began the practice of issuing commemorative coins, honoring special events such as major victories in battle (Figure 2.2). Even though these coins were still made by hand, they were relatively uniform, though not as uniform as today's machine-made coins. This non-uniformity makes ancient coins instantly recognizable.

In addition to making wonderfully artistic coins, the Greeks also made some strikingly large coins as well; albeit, these were exceedingly rare. The largest gold coin ever recovered from ancient times is a 20-stater gold piece featuring King Eucratides I of the Graeco-Bactria kingdom, measuring 5.8 cm in diameter, with a mass of 169.2 g. Only about 20 of these coins exist today. One would eventually be purchased by Napoleon Bonaparte for £1300 and now resides in the Bibliothèque Nationale de France in Paris.

Figure 2.2   Silver coin of Eucratides I (reigned 170-145 BC), a King of Bactria, with the Greek-style name in the title 'Great King'. From the British Museum's collection.
ⓒ Print Collector/Hulton Archive/Getty Images.

In most countries today, coins are produced annually, but that has not always been the case. In earlier days coins were minted on an as-needed basis—decades could go by with no new coins being issued. When the need arose lots more would then be made.

Christopher Howgego, writing in *Ancient History from Coins*, gives us an idea of what mints in the Greco-Roman world were like:

> Archaeology helps to fill in the picture. The Athenian mint from the end of the fifth century to the late first century BC has been identified at the east end of the agora. The building contained bronze bars with discs cut from similar bars, together with evidence of metal refining. Inscriptions relating to the mint were found nearby. The mint was a large establishment, 27 m by 38 m, on strong foundations 1 m thick, and consisted of a number of rooms of various sizes around an open courtyard. We happen to know that the workers in the mint were public slaves.
>
> At Rome the mint was on the Capitoline during the Republic. Remains of the building itself have been plausibly located under the church of S. Clemente, c. 400 m east of the Colosseum. It was a long rectangular building with a

width of c. 30 m, and of unknown length. It had two main storeys, the lower of which was divided into two floors in the earliest phase of the building. The exterior of the building was formed by a substantial wall, probably with only one entrance and no other openings. The lower floor was composed of a large number of rooms arranged around a courtyard with a peristyle (a row of columns with a covered roof). The complex may have continued to function as a mint until the fourth century AD. (Excerpt reproduced from ref. 2 with permission from Routledge, Copyright 1995.)

2.3 EARLY PLATED COINS

The Greeks didn't just make aesthetically pleasing coins—they were also at the forefront of numerous numismatic innovations, some of which we still use today. Most ancient coins were composed of a homogeneous alloy. Plated coins were less common, and usually represented an attempt at forgery. However, the Greeks were among the first to perfect an officially sanctioned plating process.

Analysis of a didrachm dating from 300 BC reveals a core of mostly copper with trace amounts of lead and zinc, with a coating composed of a silver-copper alloy. Careful examination of these coins by metallurgists at the University of Manchester in the UK led them to make the following conclusion:

From these observations, coupled with the examination of the whole section of the coin, it is apparent that the plating operation was effected by first making a shallow silver cup to fit the copper core, lining this with a thin sheet of the silver solder, and, after the insertion of the core, covering the whole with another inverted cup, similarly lined with solder. When the whole was afterwards reheated above the temperature of the silver-copper eutectic (alloy with the lowest possible melting point), the solder would melt and run between the two cups, fusing core and plate together. A gentle hammering of the rim would complete the preparation of the blank, which, as the twinned structure of both core and plate shows, was then struck hot. (Excerpt reproduced from ref. 3 with permission from Springer Nature, Copyright 1951.)

The Greeks made millions of coins, and since silver and gold are so unreactive, many of them are still around. At one time Athens stored millions of silver coins in the attic of the Parthenon, which served as the city-state's treasury. As paper money was not yet used, all circulating money was in the form of coins. The attic of the Parthenon was not a typical attic, but rather served as the storehouse for much of the city-state's wealth.

Ancient records claim that this storehouse could hold 10 000 talents of silver.[4] A talent would be the equivalent of 1500 silver tetradrachms, comprising a whopping 15 000 000 coins, weighing 260 metric tons. The tetradrachm was the Greek coin most commonly used for transactions—many have been found throughout the years in and around the Acropolis, the hill on which the Parthenon sits. A typical tetradrachm has a mass of around 17 g, with a volume of 1.6 cm^3. With approximate dimensions of 50 by 19 by 3 m, the Parthenon's attic had a volume of 2850 m^3, with the capacity to hold well over a billion coins! These coins were most likely housed in boxes for ease of storage and transport.

2.4 BRONZE COINS

The first bronze coins were made by the Greeks, most likely in the region of Sicily around the 5th century BC. Unlike gold and silver coins, whose face value was roughly equal to the value of the metal it contained, the face value of bronze coins was not based on their intrinsic worth (which also holds true for most coins in circulation today). These coins were known as fiduciary or token coins, and were cheaper for the government to make. Although numismatists tend to label any copper-containing coin as bronze, the composition of these cuprous coins varied widely. Some were pure copper; others contained tin, the most common additive in bronze. The importance of tin in the making of ancient coins cannot be overstated.

Tin is perfectly suited for use in alloys. It is one of the softest elements, ranking a surprisingly low 1.5 on Moh's scale of hardness, making it soft enough to be scratched by a fingernail. It is very malleable and ductile, yet it does have some heft, with a respectable density of 7.3 g mL^{-1}. Tin melts at a surprisingly low 232 °C, making it the transition metal with the lowest melting point (besides mercury). The Bronze Age owes its very existence to tin, as it was the primary element alloyed with copper. Even though tin itself is very soft, it makes copper harder when it is alloyed with it. Tin is relatively rare, however, and only makes up about 2 ppm of the earth's crust; its chief ore is casserite (SnO_2).

Tin is very familiar for its role in tin cans; the first tin cans were not pure tin, however, but were mostly steel that was tin-plated. An easy way to determine the composition of a can is to test it with a magnet—pure tin is not ferromagnetic. The popularity of tin for use in cans is owed to it being unreactive, as you do not want the can reacting with its contents; even so, tin still imparts a "tinny" taste to foods.

Aluminum foil is often erroneously referred to as tin foil, even though tin foil hasn't been in vogue since World War II. Tin is fairly stable at room temperature, but will react with oxygen to form tin oxide at higher temperatures. One of the strangest properties of tin is the eerie screeching sound it makes when a bar of it is bent, a sound known as "tin cry".

Tin would appear to be the perfect coinage metal—it is malleable, has a low melting point and is relatively inert. Alchemically, it is associated with wisdom and logic. Yet few coins are made of pure tin—its primary use has been as a component in bronze. One reason for tin's unpopularity is its unfortunate tendency to crumble at low temperatures. This behavior goes by the rather quaint name of "tin pest".

The propensity of tin to disintegrate at low temperatures has led to calamity on more than one occasion. Robert Scott's trek to the South Pole in 1912 ended tragically as the tin-solder of the kerosene cans disintegrated in the cold, causing the kerosene to leak out. Consequently, Scott and his men had no fuel with which to warm themselves; tragically all would freeze to death. Napoleon's disastrous invasion of Russia in 1812 has been blamed on the tin buttons of the soldiers' coats crumbling in the frigid temperatures, leading to thousands freezing to death, a story recounted in the wonderfully informative book that lends its name to the incident—*Napoleon's Buttons*, by Sam Kean. Even though some scholars doubt the veracity of this account, it is nonetheless feasible.

Tin has two allotropes that behave differently at different temperatures. At temperatures above 13.2 °C, tin exists in the relatively stable form known as beta or white tin. It has a tetragonal crystalline structure with a density of 7.3 $g\,mL^{-1}$, comprising the familiar form of tin. However, below 13.2 °C, this stable form of tin degrades into alpha or grey tin, with a cubic crystalline structure, having a density of only 5.8 $g\,mL^{-1}$. As alpha tin has a lower density than beta tin, the volume of tin increases as it changes from the alpha to the beta form, as its mass stays constant. It is this expansion that causes the metal to disintegrate. If any pure tin coins were made in ancient times, there is a good chance they would have all crumbled to dust by now. You can occasionally come across a tin coin, as a few countries have minted them, but they have never gained much traction.

2.5 COINS MADE FROM LEAD

Another substance often alloyed with copper is lead, which owing to its softness makes it easy to strike. Even though lead has traditionally been associated with death and darkness, there are other reasons why pure lead coins have never caught on. Lead is too soft to hold up in elemental form, and is not lustrous like gold and silver. It gets dull as it reacts with oxygen in the air. An easy way to determine the lead content of bronze coins is to measure their density—the greater the density the greater the composition of lead, as copper's density ($8.9\,g\,mL^{-1}$) is less than lead's ($11.3\,g\,mL^{-1}$).

As lead was cheap, counterfeiters would often use it to their advantage. They would make a core of lead and then add an outer layer of silver; they could often get away with this switch because lead's density of 11.3 $g\,mL^{-1}$ is pretty close to that of silver, which is 10.5 $g\,mL^{-1}$. A few Greek tetradrachms made of pure lead have been discovered—they are instantly recognizable by their dark color. A line of lead coins was produced by the kingdom of Numidia, a Roman colony, in North Africa in the second century BC, but beyond that few lead coins have been found.

The use of lead in coins has proven helpful to archaeologists, however. Analysis of lead isotopes within coins can help to pinpoint where the lead originated. Lead has a number of isotopes, the four most stable being lead-204, lead-206, lead-207 and lead-208. Each isotope has a different geographical origin. Most ancient coins will contain at least some lead, either through it being added intentionally or owing to a lack of completely purifying other metals when they were extracted from their ore, which often contains lead. By analyzing the specific ratio of isotopes within a sample of lead valuable information can be gained as to where the lead originated, or perhaps how the location of the sources has changed over time.

One method to analyze the isotopic composition of lead begins with the coins being subjected to a series of powerful laser pulses in a vacuum chamber, in a process known as laser ablation. The type of laser used is not your typical garden variety laser—laser ablation uses a Nd-YAG laser (neodymium yttrium aluminum garnet) which is used for things like eye surgery and the bombardment of tumors. These lasers can penetrate to a depth of 10 mm. The particles dislodged during these laser bombardments are then analyzed with a mass spectrometer to determine the isotopic ratios of any lead atoms present.

One fascinating study by Italian researchers involved analyzing isotopic ratios of lead-208/lead-206 and lead-207/lead-206 in 12 bronze coins from the Mediterranean region, dating from the 2nd to the 10th century AD.[5] These ratios were then compared with a database of lead mines in the region. They were able to predict with reasonable accuracy the mine from which the lead of each coin originated. Some of these mines were quite distant from where the coins were found. Archaeologists use this type of data to piece together information about trading routes, which can then be used to construct a wider picture of the region. The more recent the coin the harder it is to trace its origin, however, as there is a greater likelihood that the coins would have been melted down and reformed. If recycling of the coins happens enough times then the lead from various sources will have become so hopelessly jumbled together that it will be all but impossible to determine the origin of any of its isotopes.

Isotopic analysis has even been used to detect counterfeits. If the isotopic signature of a particular variety of coin containing lead differs significantly from those that were legitimately minted, then its authenticity is suspect.

2.6   THE ROMANS PUT THEIR STAMP ON COINS

As the Greek city-states would gradually become part of the Roman Empire, the decline in the aesthetic quality of the coinage was evident (Figure 2.3). The Romans were more practical and lacked the sophistication of the Greeks. The first Roman coins were ugly lumps of cast bronze, known as *aes grave*, or heavy bronze. Eventually respectable bronze coins would emerge from these crude forerunners. Bronze was the only metal that was possessed in abundance by the early Romans.

Coins made by the Lydians and Greeks were always struck, with only the blanks being made by casting. Some of the first Roman coins, however, were produced by casting, a rarity in the coin world. These bronze coins were so large they would have been difficult to produce any other way. As coins became smaller casting would give way to striking. The Chinese, who independently invented coinage perhaps as early as 350 BC, also relied on casting to produce coins. Most of these early Chinese cast coins are instantly recognizable by the square hole in the center (see Figure 2.4).

The Romans would learn from the Greeks however, and eventually Roman silver coins would become all but indistinguishable from Greek ones. They were even based on the same denominations, with didrachms being the first Roman silver coins made. The most common Roman coin was the denarius, which would

Figure 2.3   Ancient Roman bronze quadrans coin of Emperor Hadrian.
© Shutterstock.

Figure 2.4 Antique bronze Chinese coins.
© Shutterstock.

Figure 2.5 Silver denarius bearing the image of Julius Caesar, minted in
Rome, 44 BC.
© DeAgostini/Getty Images.

eventually be made from about 4 g of silver (Figure 2.5). A common gold coin, known as an aureus, was the same size as a denarius but much heavier (about 8 g) owing to gold's greater density, and was worth 25 denarii (Figure 2.6). The aureus would eventually be replaced with the solidus, also made of gold, which had a mass of approximately 4.5 g.

The Romans were the first to make a coin from orichalcum, an alloy of 80% copper and 20% zinc, along with trace amounts of lead, tin and other metals. It had a golden-yellow color when

Figure 2.6   Aureus (reverse), 47–48. England, Roman, Claudius I, 41–54 AD.
Gold; diameter: 2 cm.
© Sepia Times/Universal Images Group *via* Getty Images.

Figure 2.7   Ancient Roman Empire bronze coin sestertius 168 AD, laureate
head of Emperor Lucius Verus right, and left sitting female with
scales as personification of quality.
© Shutterstock.

freshly minted, making its appearance similar to that of
modern-day brass. Its name is of Greek origin and literally
means "mountain copper". Its value was second only to gold.
Most of the orichalcum that was used for coins was mined in its
current form. The most common way to make brass in Roman
times was through the cementation process.[6] Cementation
entails heating zinc ore to isolate the zinc, and then allowing
the zinc vapor to diffuse into copper to form the alloy.

Figure 2.8   Roman Dupondius Replica Coin.
© Shutterstock.

This process is difficult and time consuming, however, and was most probably not used on a large scale. The sestertius and dupondius were both made from orichalcum; the sestertius was equal to a quarter of a denarius, and the dupondius equal to half of a sestertius (see Figures 2.7 and 2.8).

Orichalcum has a long and rich history, being mentioned by Plato five times in the *Critias*, one of his later dialogues, in which he tells the story of the mythical lost island continent of Atlantis. Plato claims that the metal orichalcum was mined extensively in Atlantis—its buildings and structures were covered with this wondrous golden-colored metal. Believers in Atlantis were emboldened by news reports in 2015 that reported the recovery of 47 ingots of orichalcum from a shipwreck off the southern coast of Sicily, in just 3 m of water. Another 39 ingots were recovered from the same site in 2017. These ingots were believed to have been fashioned some 2600 years ago. They were covered in a green and blue patina, revealing the presence of copper. Analysis showed their composition to be 75–80% copper and 15–20% zinc, with trace amounts of nickel, lead and zinc, confirming that they were indeed made of orichalcum.[7]

A big difference between Roman and Greek coins was the image that adorned the coin—a typical Greek coin would feature a god or some other mythical figure. The Romans, however, popularized the practice of putting living people on coins, with Julius Caesar appearing on a number of coins. The Roman practice of depicting emperors on their coins was great propaganda, attempting to convince the citizenry that the emperor was special. The ubiquity of the emperor's image is revealed in the following account in Matthew 22:19–21 in the New Testament, in which Jesus engages in dialogue with the Pharisees, the Jewish religious leaders of the day:

> "Show Me the coin used for the poll-tax."
> And they brought him a denarius.
> And He said to them, "Whose likeness and inscription is this?"
> They said to Him, "Caesar's."
> Then He said to them, "Then render to Caesar the things that are Caesar's; and to God the things that are God's."

The legend—the text that runs around the outside perimeter of the coin—was widely used to spread political messages in Roman times. The reverse side would often contain images of the emperor's achievements or battle victories. The legend on most coins issued today contains the country that minted the coin, as well as its denomination. In the past the denomination was often not included, especially if much of the population was illiterate. Many early American colonial coins did not contain the denomination as so many people were illiterate—the denomination was evident by the size, color and imagery.

Many coins today still contain messages, often in the form of a motto—an inscription that bears a special meaning not related to the country or value. The first English language motto on US coins was "Mind Your Business", which was designed by Benjamin Franklin. In 1864, the motto "In God We Trust" began appearing on all US coins. Inscriptions on British coins have traditionally been in Latin, and often refer to God and the monarch.

The words money and mint are both of Roman origin, being derived from the name of the goddess Juno Moneta, where silver coins were manufactured near her temple beginning in 300 BC. Moneta came to be known as the goddess of money. A silver denarius from 46 BC features her image, and on the reverse are the instruments used to make money: an anvil, hammer and tongs. Her likeness still appears on money today, and was recently featured on the reverse side of a nickel-copper one-crown coin issued by the Isle of Man in 2012.

2.7 DEBASEMENT OF CURRENCY

The Romans were constantly tinkering with the precious metal content of their coins, largely to compensate for inflation and other economic woes brought on by the excesses of the emperors. As the empire declined, the precious metal content of the currency would decline as well. This debasement of coinage would continue unabated; at one point silver coins only contained 2% silver! Yet the authorities would make sure that these mostly bronze coins still contained a shiny silver surface. However, the populace was not fooled, and many would shave or melt down their more valuable older coins that had a higher silver content. Lead was often used to debase coins, with some Roman coins containing up to 30% lead. High lead content typically indicated a weak economy.

One method of debasing currency was to plate a less valuable coin with a more valuable metal. Compared to the Greeks, the Roman methods of plating were crude, and were often done with the intent to defraud. Some of the first plating attempts entailed dipping a copper coin in molten silver, or simply covering a copper coin with silver discs on both sides and then heating to fuse the metals together.

A research study undertaken by a group of French scientists attempted to replicate some of these forged coins from the third century AD, specifically some *antoniniani* (equal to 2 denarii) that were recovered in what is now northern France. The market was flooded with counterfeit coins during this time period as an assortment of amateur metallurgists set up shop in an attempt to make their own money or manipulate current money to make it worth more. If the government could debase coinage, the thinking went, then so can we.

Remarkably, the French researchers were able to produce passable replicas of these forged silver coins. They first wrapped copper discs with 50 μm thick silver foil. Next, they heated them to 950 °C in a graphite furnace for 4 minutes. The coins that emerged closely resembled known forgeries from the period. They discovered that longer heating times reduced the percentage of silver in the outer layer as copper atoms diffused from the core into this region. Heating for 4 minutes reduced the silver percentage to 26%, similar to those in the control group.

The thickness of the outer layer of these recreated coins was 200 μm, twice that of the older coins, but upon striking, the thickness was reduced by half, making their thicknesses virtually identical.[8]

Another method used by the Romans to create faux silver coins was to dip a copper blank in a solution of silver nitrate. A beautiful coating of silver would form as the copper and silver traded places. In this reaction elemental copper is oxidized into copper ions and silver ions are reduced into elemental silver. It is this pure silver that is deposited on the surface of the coin. The reaction is as follows:

$$Cu_{(s)} + AgNO_{3(aq)} \rightarrow CuNO_3 + Ag_{(s)}$$

As copper is more active than silver, it will take its place, bonding with the nitrate ion and leaving silver in the elemental form. As copper is more easily oxidized than silver, copper will more readily give up its electrons, as it is energetically favorable for copper to react with the nitrate ion. The problem with plating copper using this method is that it wears off very easily, and when exposed to air turns black. If you have a piece of copper and some silver nitrate solution, it is a very easy reaction to perform—the silver precipitates out on the copper almost immediately, forming beautiful crystals. If removed from the solution, however, the silver quickly oxidizes, forming a dark silver oxide.

As we saw earlier, gilding was commonly performed in ancient times as a way to coat base metals with a more precious metal. While generally done for decorative purposes, counterfeiters would use gilding for more nefarious reasons. Three different types of gilding processes have been uncovered on ancient coins.[9]

One type, known as foil gilding, involved wrapping a copper or silver blank with a thin layer of gold foil, which would then be hammered and heated. When struck, it produced a replica that could easily pass for pure gold.

Another type of gilding involved coating a coin blank with gold amalgam. If heated the mercury vaporized, leaving behind a pure gold coating on the coin, but some mercury would always remain. Modern-day analysis of these coins can easily detect the mercury, revealing this fraudulent technique.

Gold leaf has also been used for gilding. Gold leaf is an extremely thin film of gold that is produced by first rolling a gold ingot until it flattens out into a thin ribbon. This ribbon of gold is cut into squares and beaten until it is thin as paper. It will then be successively cut into still smaller sections and then hammered again until it is an improbably thin tenth of a micrometer, so ephemeral that the slightest air current will cause it to flutter. This thin leaf is then painstakingly applied to a coin blank, either through gentle heating or gluing. The disadvantage to forging a coin using gold leaf is that the layer is so perilously thin that it will begin to wear through quickly. Several layers of gold leaf would often need to be applied to get a coating thick enough to pass muster. If you have ever handled gold leaf you know how infuriatingly frustrating it can be to work with, as even the slightest touch can destroy it.

As gold doesn't oxidize, a "gold" coin that contains spots of corrosion is very likely a forgery, the spots being caused by areas of gold that have worn thin, revealing the underlying silver that is then oxidized into silver sulfide or silver oxide. Gilded objects of all sorts can be exposed by these telltale signs of corrosion. Three chisels on display at the British Museum in London were long believed to be made of solid gold, originating from the ancient kingdom of Ur. When the gold surface began to flake off, however, the underlying copper surface was exposed, revealing that the chisels were the product of gilding.[10]

Gilding a copper or silver object with gold, though, would be asking for trouble, as gold is almost twice as dense as either of these metals. Therefore, gilded gold coins are not that common, but gilded silver coins are. A copper coin gilded with silver would be harder to detect as their densities are so similar.

In the 3rd and 4th century AD depletion gilding was one method used to give the appearance of pure silver to a coin that was mostly copper. As copper is so much more reactive than silver, a number of acids and salts will react with copper that won't react with silver. Any of a variety of substances can be used—citric and acetic acid, and even saline, will do the trick. By repeatedly immersing a copper-silver coin in a solution that is reactive to copper but not silver, the copper will be leached away, leaving a silver coating. The silver coating produced by depletion gilding is initially porous and unattractive, but once struck is compressed, imparting a shiny silvery surface to the coin.

Depletion silvering is sometimes used to restore the luster to sterling silver jewelry. To begin the process, a piece of sterling silver is heated gently over a flame until it turns black. The black coating is copper(II) oxide, which forms as copper atoms on the surface react with oxygen. After a period of brief cooling, it is dipped into a hot solution of a weak acid. A hot solution is used as sudden temperature changes would compromise the crystalline structure of the metal. As copper(II) oxide is soluble in an acidic solution the black layer is removed. (This step is known as pickling, as it often involves the use of vinegar, the same substance used to make actual pickles.) Boric acid is another acid commonly used. This process is repeated several times—each time the black layer becomes less and less pronounced. After several cycles of heating and quenching a pristine layer of shiny silver will be present on the surface. This layer is fragile, however, and will wear off quickly. As depletion gilding is normally performed for artistic purposes, it would not be wise to try it on a family heirloom.

2.8 LOOKING INSIDE COINS

Nuclear analytical techniques are de rigueur for every branch of science these days—geologists to metallurgists have gained tremendous insight by using methods that had not been conceived of a generation ago. Archaeology is one field that has benefited immensely from our ability to manipulate subatomic particles and high energy photons into doing our bidding.

In the not too distant past, if you dug up an old coin and wanted to know what it was made of, your options were limited. You could dip it in acid to strip away the outer layer, or you could perhaps cut it in half with a pair of those amazing shears they advertise on late night infomercials. Today however, archaeologists can use a variety of non-destructive methods to look inside objects. If these techniques are used on coins, the coins have no choice but to spill their secrets.

Beginning around 540 BC, some unique silver incuse coins were struck by the Greeks in southern Italy. Incuse coins have a raised image on one side and the same depressed image on the reverse. Historians have always been curious as to the methods used to make these coins, as they are strikingly different from all other Greek coins. In 2014, Australian scientists teamed up with numismatic experts in an attempt to determine how these coins were made.[11] Neutron scattering involves bombarding a substance with beams of high-energy neutrons in an attempt to map its internal structure. As neutrons collide with atoms they are scattered, changing direction. By recording the pattern of scattering, a diffraction pattern can be constructed, yielding a picture of the internal arrangement of atoms within a substance.

Neutron scattering provides a way to determine if the core of a coin is made from a different substance than the rest of the coin. In the Australian study, one allegedly pure silver coin was actually discovered to have a bronze core; this determination was only made possible because neutrons encountering bronze are diffracted differently than those encountering silver. Also, if metals are subjected to hammering or other types of stresses, the crystal orientation of the metal atoms changes, creating a distinct pattern that can be detected by neutron diffraction. Coins made during a different period will have a different internal structure based on the technologies of the era—neutron diffraction can reveal this internal fingerprint. Data obtained from neutron scattering technology revealed that these incused coins were made from a forging process in which the metal was deformed by hammering, but not at the extremely elevated temperatures that were typical for the time.

Recent studies involving the use of X-ray fluorescence have yielded exciting findings in determining the composition of ancient Greek and Roman coins. X-ray fluorescence is a nondestructive technique that involves bombarding a coin with high energy X-rays. There are two broad categories of X-ray fluorescence: energy dispersive and wavelength dispersive.

With energy dispersive X-ray fluorescence, coins can be X-rayed without being cleaned. Just like you can be X-rayed with your clothes on, coins can also be X-rayed through a layer of crud, provided the corrosion layer is not too thick. The ionizing radiation of the X-rays is of sufficient energy to dislodge an electron from an inner orbital atom within the sample being tested. In an attempt to become stable, an electron from an orbital in a higher energy level will fall into the gap left by the exiting electron. In dropping down into this lower energy state, secondary or fluorescent X-ray radiation is emitted—the amount of emitted radiation being equal to the difference in energy states between the initial and final positions of the electron. By measuring the intensity of this emitted radiation, the exact composition of the sample can be determined as different elements release differing amounts of energy, depending on the specific arrangement of their electrons. This method can be used to determine the chemical composition of everything from cement to glass to metal ores.

One study performed by scientists from the Institute of Earth Sciences Jaume Almera in Barcelona analyzed 132 Greek drachmae from the fourth to the first century BC in an attempt to ascertain their composition.[12] Analysis of the data showed some interesting patterns. The percentage of silver held steady at above 98% throughout much of this period, but during the 260–220 BC span it dipped below 98%, indicating an economic downturn as cheaper metals were substituted for more precious ones.

X-ray fluorescence is useful for more than just determining the exact composition of coins. It can also be used to determine the extent of surface oxidation. Older coins will have significantly greater amounts of surface oxidation, sometimes reaching hundreds of microns thick, as opposed to newer coins that may have significantly less corrosion.

Wavelength dispersive X-ray fluorescence is more limiting than energy dispersive, in that it can only penetrate the outermost layer of the coin, to a thickness of about 5–10 μm, therefore its use requires careful cleaning of the coin to remove any surface oxidation.[13] In the sample to be tested, all of the different elements will be excited simultaneously as they are bombarded by X-rays. The emitted radiation is focused through a crystal that diffracts different wavelengths into different directions. By analyzing the diffraction pattern, the chemical composition of a sample can be determined.

Chemists from Eastern Michigan University, USA, analyzed 22 copper-based Roman coins dating from 217 to 31 BC in an attempt to ascertain their composition, specifically the amount of iron, cobalt, nickel, copper, zinc, arsenic, silver, tin, antimony and lead within the coins. They discovered a higher concentration of iron in the brass coins than in the copper or bronze coins. They were especially impressed with the relatively high amount of cobalt found in a number of the coins.

Cobalt is rarely found in the newer Roman coins, but was quite common before 135 BC, owing to the source of ore used to make the coins. They discovered that the amount of cobalt could be a useful way to date Roman coins. When they discovered that a newer Roman coin had higher than expected levels of cobalt, closer examination revealed that it had been struck over a much older coin, a common practice at the time. Using X-ray fluorescence, they were able to determine the exact percentage within the coins of each of the 10 elements under study. From their analysis they were able to draw several conclusions about Roman coinage. They discovered that widespread remelting of coins did not occur for the time periods they studied, as evidenced by the markedly different chemical compositions of older coins as opposed to newer ones. In addition to being high in cobalt, they also learned that earlier coins had a higher lead, arsenic and antimony content than previously thought.[14]

As copper and silver differ in density it might seem as though a simple density determination would be sufficient to detect a genuine silver coin from one that is merely silver plated. However, owing to the irregular shapes and unevenness of many ancient coins, some silver-plated coins can actually have a higher density than a pure silver coin. Corrosion products can also throw off density calculations. Therefore, density alone is seldom sufficient to determine the exact composition of a coin.

One method used to discriminate between genuine silver coins and plated forgeries involves bombarding the coins with thermal neutrons to determine their composition. A thermal neutron is a neutron that is not bound to the nucleus of an atom but is emitted by a radioactive source and then slowed to room temperature by a moderator. (The neutrons are termed *thermal* because they are in thermal equilibrium with their surroundings.)

When bombarded by neutrons silver is more opaque than copper, as silver is a much better neutron absorber than copper. One study used this principle to determine the authenticity of a number of ancient Greek and Roman coins dating from the 3rd century BC to the 1st century AD. A pure silver coin placed between the neutron source (a sample of radioactive Cf-252) and the neutron detector absorbs more neutrons than a silver-plated coin, which would be predominantly copper. Over 300 denarii, all of which appeared to be pure silver, were tested. Thermal absorption revealed that 12% of them were actually silver-plated.[15]

2.9   MEDIEVAL COINS

Even though kingdoms rise and fall, their coins remain. After the fall of the Western Roman Empire in the 5th century AD, the Eastern empire lived on, as the Byzantine Empire, which lasted for another 1000 years. The production of coins continued during this era, as did the circulation of the untold millions of coins produced during the days of the Roman Empire. The silver denarius was still the most popular coin in the early Middle Ages.

The modern English system of currency had its roots in the Early Medieval Period, with currency reforms instituted in the 8th century by Pepin the Short, the father of Charlemagne. Pepin introduced the units of pennies, shillings and pounds. Each silver penny was to have a mass of about 2 g, with 11 pennies equal to 1 shilling, and 22 shillings equal to 1 pound. A pound of pennies actually weighed a little more than an actual pound, but was fairly close. This arrangement would later be modified so that 1 pound would be equal to 20 shillings, and 1 shilling equal to 12 pennies (or pence). This system would last until 1971.

Despite the various denominations introduced by Pepin, the vast majority of new coins made in Europe for the next 500 years would be silver pennies. A shilling and a pound were just book keeping measures that made it easier to keep track of your money. Eventually gold and copper coins would be introduced, and a few others.

Archaeologists of the future will have little trouble dating modern coins as the year is prominently displayed. Dates were not a feature though on ancient coins. The first date did not appear on a European coin until the Middle Ages, with a date of MCCXXXIIII (1234) appearing on the reverse side of a silver penny minted in Denmark.

Although parts of the Medieval Period are sometimes referred to as the Dark Ages, a number of scholars dislike this term, as many great works of art were produced and scientific discoveries made during this time. The scientific method, the evidence-based form of reasoning that is so taken for granted today was developed during this time, being based largely on the work of Thomas Aquinas (1225–1274) and Roger Bacon (1214–1292), among others. The work of the alchemists continued unabated, and while they never did find the philosopher's stone, they managed to isolate and describe a whole host of substances that no chemistry lab could do without—substances such as hydrochloric, sulfuric and nitric acid as well as alkaline substances like sodium carbonate, potash and ammonia. The only reason the Renaissance appeared to burst on the scene was because the seeds had been planted and were slowly growing for so many years prior.

Perhaps the biggest impediment to progress during the Middle Ages was the Black Death of the 14th century which would kill 100 million people—60% of Europe's population. Doctors had no answers and no way to stem the epidemic, having no idea that bubonic plague was transmitted by bacteria (*Yersnia pestis*), which in turn were spread through flea and lice bites. They knew enough to control rats and other rodents that carried fleas, but beyond that they often relied on folk remedies to stem the disease, such as a pocketful of posies.

However, one method of prevention did have some merit, even though its overall effectiveness was minimal at best. To prevent the spread of the plague, townspeople tried to limit interaction with anyone who might be a carrier. To minimize contact, it was common practice to place coins in holes that had been bored into rocks along the periphery of the town. Vinegar would then be poured into the holes to disinfect the coins. Merchants would leave goods at the entrance to the town and accept the coins as payment. Copper coins were commonly used in those days; the vinegar would react with the copper to create copper acetate, the same substance Julia Child used as a shark repellant for the Navy. A related compound—copper(II) acetate triarsenite, is the active ingredient in Paris Green, which has long been used for rat poison, being a very effective rodenticide and insecticide.

Even though many of the same techniques used by the Romans were also used during the Middle Ages, there were significant innovations. One difference is that medieval coin makers tended to use coin blanks that were cold, relying on sturdy coin dies made of hardened iron that could churn out 20 000 coins per day. A number of beautiful coins of all shapes and sizes were produced during this period.

The imagery of the coins took on a distinctly Christian flavor, with depictions of Jesus and the cross commonly appearing. However, other coins were decidedly unartistic, with crude lettering and wooden-looking portraits. The copper coins were especially crude, but were deemed good enough for the illiterate serfs. The nummus, the forerunner of the copper penny, was produced in enormous quantities. This bronze coin was only 10 mm in diameter and had a mass of just half a gram. Gold and silver coins, which were more likely to be possessed by the nobles, were much more artistically rendered. The coins of the Middle Ages clearly reflected the underlying class structures of European society during this time.

Some of these gold coins would lay dormant for centuries. While excavating for a hotel in Pisa, Italy in 1925, workers discovered a horde of gold coins, 229 in all, dating from the 13th century. A number of these coins were analyzed using X-ray fluorescence in an attempt to discover their composition.[16] Among the coins tested were a number of florins, beautiful gold coins first issued by the Republic of Florence in 1252, serving as gold bullion throughout much of Europe. Their gold content averaged 99%, with trace amounts of iron and lead. The high purity of gold used in the florin showed that the economy of Florence was likely thriving during this time.

Other coins from this horde were shown to have a lesser concentration of gold. The tari, a much smaller gold coin that was widely circulated, had an average gold concentration of 72%, along with 22% silver and trace amounts of copper, iron, palladium and lead.

One interesting feature of coins from the Byzantine era was their tendency to get progressively thinner. They were so thin that blanks could be cut from sheets of metal with heavy duty shears. Some coins from this period were actually more square than round, showing that they had been sheared rather than punched out. Many gold and silver coins were so thin they could be bent by hand. To counteract the inherent weakness of these thin coins, many coins of the late Byzantine Empire were distinctly cup-shaped. To strike such coins, the lower die had a greater curvature than the upper die. The curvature of these coins tended to prevent the edges from the very thin coins from cracking, as a dome-shaped structure is inherently strong, effectively distributing compressive forces that might act upon it.

Although rarely seen today, some modern coins are cup-shaped. In 2009, France issued two cupped coins—one commemorating the 40th anniversary of the moon landing and the other recognizing the International Year of Astronomy. In 2014 the US mint released some cup-shaped gold and silver coins commemorating the 10th anniversary of the Baseball Hall of Fame. These unique coins feature a baseball mitt on the concave observe side, while the convex reverse side is designed to resemble a baseball, complete with stitching.

2.10   MECHANIZATION

Although centuries of coin making led to improvements in efficiency that would lead to more uniform-shaped coins and more intricate designs, the basics of coin-making throughout much of history were not that different from what was used to strike the first coins in Lydia. Making coins was still a labor-intensive process that was performed by hand. Even the best of these hand struck coins were crude when compared to modern coins, but coins were about to get a makeover.

Leonardo da Vinci drew up plans for a coin press around 1500, though he never actually assembled it, as far as we know. Similar plans would come to fruition in 1550, when a very efficient coin press was developed by the German metalworker Marx Schwab, whose coin press could make coins of uniform thickness. A coin press is simply a machine that can press an image onto a coin blank faster and better and with more precision than can be done by hand.

The coin press consisted of an anvil die on the bottom and a pile die on top, which was attached to a large screw that in turn was attached to the center of a large rod known as the balance arm—the longer the arm, the greater the mechanical advantage. One man would place the coins between the dies, and either two or four men, known as spinners, would operate the handle, which would turn the screw and strike the coins. Although the coin press was very efficient, it would be another 100 years before it was widely used, owing to opposition from traditional coin makers who feared losing their jobs. However, as these machine-made, or milled coins, were very difficult to counterfeit, it was only a matter of time before every mint in the world would be using a coin press.

One of the best records of what it was like to make coins using one of these new screw presses was furnished by Samuel Pepys, a government official and Parliament member who is best remembered for his diary. He kept a detailed record of events from 1660–1669, furnishing a vivid account of life and events during that decade. His diary would go on to be published in 1825, and is widely considered to be the most influential diary ever written. Fortunately for us, he took a tour of the Royal Mint on May 19, 1663, not long after the screw press became a permanent fixture there, and wrote about it, giving us a 10-step description of the steps needed to make money in the 17th century:

1. Before they do anything they assay the bullion, which is done, if it be gold, by taking an equal weight of that and of silver, of each a small weight, which they reckon to be six ounces or half a pound troy; this they wrap up within lead. If it be silver, they put such a quantity of that alone and wrap it up in lead, and then putting them into little earthen cupps made of stuff like tobacco pipes, and put them into a burning hot furnace, where, after a while, the whole body is melted, and at last the lead in both is sunk into the body of the cupp, which carries away all the copper or dross with it, and left the pure gold and silver embodyed together, of that which hath both been put into the cupp together, and the silver alone in these where it was put alone in the leaden case.

After separating out the impurities from the gold and silver, Pepys next describes how to separate the gold from the silver.

Aqua fortis is the ancient name for nitric acid, which is Latin for strong water. It reacts with silver but not gold, producing a solution of silver nitrate, which would have looked like water:

> And to part the silver and the gold in the first experiment, they put the mixed body into a glass of aqua-fortis, which separates them by spitting out the silver into such small parts that you cannot tell what it becomes, but turns into the very water and leaves the gold at the bottom clear of itself, with the silver wholly spit out, and yet the gold in the form that it was doubled together in when it was a mixed body of gold and silver, which is a great mystery; and after all this is done to get the silver together out of the water is as strange. But the nature of the assay is thus: the piece of gold that goes into the furnace twelve ounces, if it comes out again eleven ounces, and the piece of silver which goes in twelve and comes out again eleven and two pennyweight, are just of the alloy of the standard of England.

Once they isolated the pure metal, the process was quite similar to how modern-day coins are made, as we shall discuss a little later. The molten metal is poured into molds and then rolled and flattened into a strip equal in thickness to that of the coin being made. From this strip the blanks were cut:

> 2. They melt it into long plates, which, if the mould do take ayre, then the plate is not of an equal heaviness in every part of it, as it often falls out.
> 3. They draw these plates between rollers to bring them to an even thickness all along and every plate of the same thickness, and it is very strange how the drawing it twice easily between the rollers will make it as hot as fire, yet cannot touch it.
> 4. They bring it to another pair of rollers, which they call adjusting it, which bring it to a greater exactness in its thickness than the first could be.
> 5. They cut them into round pieces, which they do with the greatest ease, speed, and exactness in the world.

The whole point of the modern coin-making machinery was to standardize production, so that all coins of the same composition had an identical mass, volume and shape. Pepys goes on to discuss how the coin makers met these specifications:

6. They weigh these, and where they find any to be too heavy they file them, which they call sizeing them; or light, they lay them by, which is very seldom, but they are of a most exact weight, but however, in the melting, all parts by some accident not being close alike, now and then a difference will be, and, this filing being done, there shall not be any imaginable difference almost between the weight of forty of these against another forty chosen by chance out of all their heaps.

7. These round pieces having been cut out of the plates, which in passing the rollers are bent, they are sometimes a little crooked or swelling out or sinking in, and therefore they have a way of clapping 100 or 2 together into an engine, which with a screw presses them so hard that they come out as flat as is possible.

Pepys only gives one line to the next step, being uncharacteristically terse:

8. They blanch them.

Blanching is a vitally important step, and refers to whitening of the coin blanks. Any process that makes something whiter is known as blanching, which derives its name from the French word *blanc* for white. Blanching can refer to any process that produces a lighter color. If you press down on your skin hard enough to turn it white you will have blanched your skin. During the reign of King Henry VIII in the 16th century, a common method to debase currency was to blanch copper blanks with silver. This coating of silver was so thin that it wore off on silver coins bearing the king's image, usually wearing off on the highest point of relief, which happened to be the nose, earning him the dubious moniker of "Old Coppernose". However, this form of debasement had stopped by Pepys's time, so blanching here referred to the process of dipping the blanks in an acidic solution, most likely alum, to remove the dark layer of surface oxidation during annealing, where the coins were heated to soften them up in anticipation of being struck. After blanching the coins would be dried with sawdust.

Next Pepys refers to a secret process pioneered by Peter Blondeau, an engineer on loan from the Paris mint, of producing inscriptions along the edges of coins in order to prevent coin clipping, the practice of shaving off the edges of coins to retrieve some of the precious metal, a widespread (but illegal) practice at the time. This device consisted of a metal bar firmly fastened to a table, with a movable metal bar a short distance away held in place by sturdy springs and attached to a crank. The gap between the two bars was just less than the circumference of the coin. A negative image of the inscription would be engraved onto one or both of the bars, and when the crank was turned the inscription would be impressed onto the edge of a coin:

9.   They mark the letters on the edges, which is kept as the great secret by Blondeau, who was not in the way, and so I did not speak with him today.

Finally, Pepys refers to the speed with which these new coin presses could mill coins:

10.   They mill them, that is, put on the marks on both sides at once with great exactness and speed, and then the money is perfect. The mill is after this manner: one of the dyes, which has one side of the piece cut, is fastened to a thing fixed below, and the other dye (and they tell me a payre of dyes will last the marking of £10 000 before it be worn out, they and all other their tools being made of hardened steel) to an engine above, which is moveable by a screw, which is pulled by men; and then a piece being clapped by one sitting below between the two dyes, when they meet the impression is set, and then the man with his finger strikes off the piece and claps another in, and then the other men they pull again and that is marked, and then another and

another with great speed. They say that this way is more charge to the King than the old way, but it is neater, freer from clipping or counterfeiting, the putting of the words upon the edges being not to be done (though counterfeited) without an engine of the charge and noise that no counterfeit will be at or venture upon, and it employs as many men as the old and speedier. They now coyne between £16 and £24 000 in a week.[17]

2.11 ISAAC NEWTON

Against this backdrop entered Isaac Newton, who, among other things, was quite the alchemist. According to some scholars, alchemy was more than a pleasant diversion for Newton, but was rather his all-consuming passion, his one true love. Not only did he devote enormous brainpower in attempting to uncover alchemical mysteries, but he enjoyed getting his hands dirty, not relying on aides to do the grunt work involved in running a laboratory, instead taking it upon himself to run countless chemical tests, keeping meticulous records of every step involved. Newton even constructed his own furnaces in which to perform his reactions.[18]

Of the 10 to 15 million words Newton is believed to have written, 1 million are solely on the subject of alchemy.[19] Upon his death he was found to have had in his possession 170 books on alchemy.[20] In his writings Newton often alluded to being tantalizing close to finding the famed philosopher's stone, which would have made his years of toil worth it. In 2016 The Chemical Heritage Foundation in Philadelphia acquired a manuscript at auction in Newton's own handwriting, entitled "Preparation of the Mercury for the Stone by the Antimonial Stellate Regulus of Mars and Luna from the Manuscripts of the American Philosopher". The recipe was copied from one devised by George Starkey, a Boston area doctor who was also a prominent alchemist. The manuscript contained a detailed procedure for synthesizing "sophick" (or philosophic) mercury. This sophick mercury was allegedly an important step to making the philosopher's stone, which could then be used to convert any base metal into gold.

It is a matter of historical record that Newton suffered a breakdown of sorts from 1692–1693, undergoing severe depression, or perhaps madness, barely leaving his house for months at a time. It was no secret that Newton was difficult to be around—he had few friends and his volatile personality was the stuff of legend. He has been posthumously diagnosed as suffering from everything from Asperger's to bipolar disorder. His nervous breakdown has at times been attributed to mercury and lead poisoning from all of his alchemical experiments. Elevated levels of both mercury and lead found in samples of Newton's hair give some credence to this hypothesis.[20] However, as his breakdown came on the heels of claims that he was tantalizingly close to discovering the philosopher's stone, he may have been disillusioned as he was forced to realize his aspirations would never come to fruition. After his depressive episode ended, Newton would never pursue alchemy with the same vigor as before.

When Newton was offered the position as Warden of the Royal Mint in London in 1696, he readily accepted, eager to leave the backwater environs of Trinity College in Cambridge for the excitement of the sprawling metropolis that was London; with a population of 600 000 it was by far England's largest city. Newton would soon discover that the many years he spent measuring and mixing chemicals was not a waste of his time after all.

When Newton accepted his position England's monetary system was in shambles. One in ten coins in circulation was counterfeit. Those that were genuine were often so frequently clipped that they were worth only a fraction of their face value. If a silver coin was whole, the silver itself would typically be worth more than the face value of the coin. As a result, money was rapidly leaving England for France or the Netherlands, where the silver would be melted down and traded for a profit. The acute shortage of physical money threatened to destroy the economy. There were also two types of money rattling around—the old hand-struck coins and the new machine-made coins. As these new milled coins couldn't be clipped they still contained their full weight of silver—those that weren't melted down for their silver content were frequently hoarded.

Newton threw himself into his new position. He would eventually become Master of the mint, spending the last 30 years of his life as a civil servant. Many of his duties were logistical in nature. He oversaw the Great Recoinage of 1696, during which the old hand-hammered silver coins were got rid of once and for all. As they were no longer accepted as payment for taxes, most Britons jumped at the government's offer to buy them for five shillings and eight pence per ounce, a fair price. They were then melted down and refashioned into modern milled coins. Newton ordered new furnaces, rolling mills and coining presses to get the job done. In no time the mint was churning out a £100 000 per week.

Newton pursued counterfeiters with surprising vigor, and even interrogated suspects. Despite his many personal foibles, no one ever doubted Newton's integrity, and he set about rooting out corruption and incompetence from the Mint. When a set of coining dies went missing, Newton was on the case. One of his responsibilities was that of ensuring the gold used for coins was up to legal standards. The amount of gold in minted coins was subject to inspection, and for gold coins was required to be 22 karats, or 916.6 parts gold per 1000 parts total, ± 3.5 parts. The remainder was to be silver and copper. This 22-karat gold, also known as crown gold, had been the standard since the days of King Henry VIII, who changed it from the softer 23-karat gold, as the harder 22-karat gold was more difficult to clip. As Newton oversaw the refining and assaying of gold for the Mint, it was up to him to make sure it was of sufficient purity. In his many previous experiments, Newton had discovered antimony to be an extremely useful substance in helping to separate pure metals from their ores.[21] He suggested that antimony be used at the Mint to attain a greater purity than aqua fortis itself might provide.

The use of antimony sulfide (Sb_2S_3) as a way to separate gold from its ores has been used for centuries by alchemists. When a melted alloy is added to antimony sulfide, other metals such as copper or silver will quickly react with the sulfur while antimony will form a complex with gold. These sulfide compounds are less dense and will rise to the top of the mixture, where they can then be easily skimmed off. If the gold-antimony complex is heated strongly in an atmosphere of oxygen, antimony trioxide (Sb_2O_3) will form, leaving behind pure gold.

In one 1707 inquiry, inspectors determined that the trial gold plate presented by Newton was outside of acceptable tolerances by 1 part per 1000. It was actually finer gold than what was required, which was therefore costing the Mint money. Newton was none too happy with this criticism of his work, and true to form offered his own scathing criticism of the inspectors and their shoddy methods. He subsequently tested his plate and reported it to have a fineness of 921.0, which was indeed over the acceptable limits. It appeared as if Newton had done his job too well. The matter was dropped—after all, who was going to quarrel with Isaac Newton? Over the years Newton's plates would be tested again, as recently as 1974 by two separate assayers. They came up with values of 919.8 and 919.9, not as pure as Newton claimed, but actually within the acceptable tolerance levels.[21]

2.12 THE WONDERS OF STEAM

Coin making would change again with the advent of the Industrial Revolution, which was ushered in by steam, of all things. There is perhaps no substance on earth as misidentified as steam. For starters, it is invisible. Anytime someone claims to see steam or points out a visible substance as steam, they are mistaken, for what they are really seeing is white wisps of liquid water resulting from the condensation of steam. When water boils in a tea kettle the white cloud above the whistling spout is liquid water—not steam—that forms when the steam condenses upon encountering the cooler surrounding air. If you apply a blowtorch to this white cloud you can boil the liquid water back into steam. If you look carefully at the spout of a teakettle, an area a half-centimeter or so high is readily visible in which there is no visible cloud—this area of invisibility represents the steam.

Steam in an enclosed container is capable of exerting great pressure, hence all of the safety precautions that must be followed when using a pressure cooker. The idea of harnessing this pressure and putting it to work would not only usher in the Industrial Revolution, it would also forever change the face of coin production.

The steam engine, which originated with English inventor Thomas Newcomen in 1712, was the vehicle by which steam would be exploited. His device was elegantly simple. By pumping steam into a cylinder that was fitted with a piston, a region of reduced pressure (sometimes known by the oxymoronic expression "partial vacuum") could be created by spraying the cylinder with cold water, condensing the steam. Atmospheric pressure would then force the piston into the cylinder. It was the continuous up and down movement of the piston that creates useful work.

A simple demonstration can vividly illustrate how a steam engine works. A little water is first put in the bottom of an aluminum soda can and then heated to boiling. If this can of boiling water is quickly inverted into a container of cold water the can will instantly implode. As the can of boiling water is essentially a can full of steam, it condenses when it hits the cold water, creating a vacuum, allowing outside air pressure to crush the can.

James Watt, a Scottish engineer and chemist, would later improve on Newcomen's design and thus garner the lion's share of the credit for the invention of the steam engine. Watt's main improvement was having a separate chamber connected to the cylinder housing the piston where steam condensation occurred and a vacuum was created. This innovation greatly improved the efficiency of the steam engine, making it a force to be reckoned with. However, it was Watt's partnership with Matthew Boulton, an English businessman, which would further increase the efficiency of steam power.

The modern era of coin making began in 1789 with Matthew Boulton in Birmingham, England. Counterfeiting was rampant—two-thirds of the coins in circulation at that time were probably counterfeit. Every new innovation, it seemed, could be copied. Producing coins that were of uniform size, shape and weight was more important than ever. It was also necessary to make the designs on the coins increasingly intricate so they would be even more difficult to replicate. Coin blanks would still be cut out of sheet metal, put into the press, and stamped with an engraved die, but now each of these steps could be done faster and better than before.

However, there was one innovation that Boulton's steam-powered coin presses automated that stood above all others, which would become the distinctive feature of modern coins. The blanks were restrained by a metal collar that prevented the coin from widening when stamped. A raised rim was thus created in the coin—a feature that would be very difficult to replicate, and to date the best anti-counterfeiting measure ever devised for coins. The biggest difference between modern coins and older coins is this raised rim—older coins are usually completely flat.

These early presses could strike up to 84 coins per minute. The first coins were made for the East India Company; the first British coins produced by steam power were minted in 1797— the one-ounce penny and the two-ounce two pence. They were so uniform that the diameter of 17 pennies laid out would equal exactly 2 feet; eight two pence would equal one foot. They were the first copper pennies ever made, the face value of the coins being based on the price of copper at the time. They were beautiful coins, becoming affectionately known as cartwheels (Figure 2.9). However, since they were made from pure copper, they tended to dent easily. About 45 million were produced, but only for a short time—they would be discontinued by the end of the century. The copper penny would eventually give way to bronze, an alloy that was much more durable than pure copper. However the raised rim and incuse (indented) inscriptions of these early pennies proved extremely difficult to counterfeit. It wasn't long before the majority of the world's coins were produced by steam-powered presses.

The restraining collar on the modern coin presses made it possible for the first time to make coins with reeded edges—an

Figure 2.9 British cartwheel penny.
© Shutterstock.

innovation that has lasted to this present day. A coin with a reeded edge contains many small vertical grooves created during the striking process. Grooves etched on the inside of the collar are impressed on the edges of the coin when the coin is struck. Although the reeded edges were designed to make coin clipping more difficult, today they serve more of an aesthetic purpose.

American dimes have 118 ridges, quarters have 119, and pennies and nickels have none. The 1-euro coin has a reeded edge that alternates with a smooth edge. Australian 1 and 2-dollar coins also contain this interrupted pattern of reeding. There are many other variations on the edge. The 2-cent euro coin has a deep groove running along its entire circumference, giving it the appearance of having been made by sticking together two separate coins. One unintended advantage of the ridges is that it helps the visually impaired to distinguish between different types of coins.

Inscriptions still appear on the edges of coins, which not only increase their aesthetic appeal, but represent an added security feature. Early American half and silver dollars had their value etched on the edge of the coins. Modern edge inscriptions typically involve a separate step, and are imprinted onto the coin blank before the face of the coin is struck. The 1997 British two-pound coin honored Isaac Newton with the quote, "STANDING ON THE SHOULDERS OF GIANTS," inscribed on its edge. Interestingly, the inscription may be facing in different directions on different coins owing to the speed with which the coins are struck, as it would not be efficient to have all the blanks with the edge inscription initially orientated the same way.

Eventually steam power would be supplanted by electricity. Today, the Royal Mint produces 2 billion coins per year for all of the United Kingdom. In 2014 the US Mint in Philadelphia produced a record 42.44 million coins in a single day. These staggering numbers are only possible owing to ever-increasing automation.

2.13   THE FIERY FURNACE—WHERE ALL COINS BEGIN

If you look at modern coins, you will quickly discover two main categories: coins that are composed of a homogeneous alloy, and those that are heterogeneous, or layered. The making of both types begins the same way—in a fiery furnace. Just like in ancient times, various elements are combined in specific quantities and then heated in a furnace to form an alloy with the specific desired properties of hardness, durability and corrosion resistance.

Most people think of a furnace as the device that heats your home, the thing connected to the thermostat. However, a woodstove is also a furnace, as is the crucible you use to heat substances in the chemistry lab. When I was a kid we lived in a rural area, and most homes had a burning barrel out back where you would burn your combustible rubbish. That was a furnace. A furnace is any container that holds burning substances, being based on the Latin word *fornax*, which means oven or kiln. So, if you are baking a pie and you burn it, then yes, your oven becomes a furnace. In a household gas furnace, there is of course a literal fire inside the furnace as the natural gas combusts. If you heat your home with electric there is of course no fire, and thus no furnace. Even a cylinder in the internal combustion engine of an automobile is a furnace, as that is where the fuel burns.

If you peer at molten metal in a furnace, the first thing you will notice, other than the intense heat being radiated, is the red/orange/yellow glow color. Few sights are more impressive than glowing metals. Metals heated in a furnace always emit visible light—the color depends on the temperature. This emission of visible light owing to heating is incandescence. The incandescent bulb utilizes this principle as electrons pass through the very thin tungsten filament within the bulb, generating heat, which in turn generates light.

Any object above absolute zero will emit electromagnetic radiation in some form, usually in the infrared portion of the spectrum. However, at about 500 °C, a dull red color will be evident, which is normally referred to as "red-hot". At 1700 °C an orange color can be observed, such as in a candle flame when unburned bits of carbon incandesce. At 2500 °C the yellow-white color of an incandescent light bulb can be seen. Our sun, at 5800 °C, is white, while some stars at 6000 °C are blue-white, the hottest color of incandescence. The blinding white color of burning magnesium, or an old-fashioned camera flashbulb, is due to incandescence.

Incandescence is not burning, as a light bulb filament does not burn but rather glows. If the glass globe is removed from an incandescent light bulb and the filament is left intact, it will burn out in about a second when turned on—the tungsten filament combines with oxygen and literally burns in half, forming white powdery tungsten oxide all over the filament. When a light bulb "burns out" from typical use the filament breaks in half owing to thinning from sublimation of the tungsten atoms, which are then deposited on the inner surface of the bulb, forming a black spot. You can easily observe the incandescence of any coin by holding it in the flame of a Bunsen burner or propane torch. In a short time, it will glow a beautiful red color.

Furnaces used to melt metal are a special breed of furnace, as they get exceedingly hot. Most metal melting furnaces are found in foundries, which are factories in which metals are cast. Some mints, such as the Royal Mint, have their own foundries; the US Mint outsources their foundry work, receiving premade strips of metal ready to be made into coins.

Both the furnace and the molds must be made of heat resistant material that can withstand these tremendously hot temperatures. In the old days stone or brick were used, which are good insulators. The insulting material in a furnace is known as a refractory; the material of choice has typically been ceramics, which can withstand incredibly high temperatures. Porcelain is perhaps the most well-known example of refractory material, which itself is made by firing up clay to high temperatures in a kiln. Ceramics differ from glass in that ceramics are opaque and possess some type of crystalline structure, whereas glass is amorphous.

2.14 SHAPING THE METAL

Just like in the old days, modern coin making always involves casting. The process begins when solid ingots of whatever metal is desired are cast into a charging furnace until they are completely melted. At any foundry, samples of molten metal will be periodically taken from the furnace and analyzed to make sure the composition is acceptable, with X-ray fluorescence being one commonly used analytical technique. Once molten, the metal is transferred into a holding furnace until it is ready to be poured. Holding furnaces do not need to be as hot as a charging or melting furnace, so using them cuts down significantly on energy costs. The molten metal is then poured into a long rectangular mold so it can cool into a thin rectangular block. At the Royal Mint, the mold is placed between two slabs of water-cooled graphite to facilitate hardening. Graphite can withstand incredibly high temperatures, and is often incorporated into the refractory materials. At the Perth Mint in Australia visitors can watch precious metals being poured into a crucible of clay and graphite that is capable of withstanding temperatures up to 1600 °C.

Oxidation occurs rapidly at such elevated temperatures, therefore once this molten metal hardens into a thin rectangular block it will be coated with a layer of black oxide. Whenever metals are heated they react more quickly with oxygen and other gases in the atmosphere. You will probably notice that anytime a metal is heated it will look different afterwards—at high temperatures the reaction rate between the metal and the surrounding hot air greatly facilitates oxidation on its surface, as there are many more collisions between the metal and surrounding gaseous atoms. If you heat any coin over a flame and then douse it in cold water, you can clearly see this ugly layer of tarnish. The corrosion layer of the rectangular block is scraped off as it passes through lubricated rollers, leaving behind a clean, glistening surface.

Next, the malleability of metals is put on full display as the block is subjected to enormous compressive forces. The block is positioned between heavy duty rollers that flatten it into a strip—when exactly it goes from a block to a strip is perhaps a matter of philosophical conjecture. The process is repeated over and over as the rollers come ever closer together, until the strip is the precise thickness of the desired coin. You may have seen these types of rollers in action if you have ever received one of those flattened oval souvenir pennies at an amusement park. Once the strip is flattened to the desired thickness it is wound up in a giant coil, where it awaits the next step.

All coins minted today begin their life from one of these giant rolls of sheet metal. The US mint purchases giant coils that are premade—each is 33 cm wide and over 500 meters long, weighing in at around 3 tons. A blanking press will then punch little discs of metal out of this sheet like a cookie cutter. An efficient blanking press will punch out 20 blanks at a time, completing five up-and-down cycles each second. In one minute, 6000 blanks can be punched out. Sometimes the blanking press doesn't work properly and coins can be clipped off or otherwise distorted owing to the strip not feeding properly, or perhaps owing to problems with the blanking die itself. It is extremely rare for a coin with a blanking error to make it out of the mint— those that do are highly sought after by collectors. After punching, the leftover metal (known as the webbing) is recycled and put back into the furnace.

2.15 GETTING THE COINS READY—ANNEALING AND PICKLING

The next step in making coins involves an annealing process. To anneal a metal, it must go through another furnace and again be heated to a high temperature, but not high enough to melt it. The coins may be placed on a conveyer belt for this step. This process typically takes anywhere from 45–90 minutes, depending on the type of metal; nickel tends to take the longest to anneal. It is crucial that the metal be allowed to cool slowly after annealing, however. A metal that has been annealed will be soft and malleable, which can then be easily struck. The opposite of an annealed metal is a hardened metal, which is formed when a metal is heated and then cooled by rapid quenching in cold water. The resulting metal will be hard, yet brittle. It would be difficult to strike a coin blank that has been hardened.

Modern British pennies are made of steel, which can be difficult to deal with. Strips of steel are not made by the Royal Mint. As steel has a much higher melting point than most of the alloys used to make coins, specialized furnaces and equipment are needed to make it. Therefore, this step is outsourced and coils of steel are brought in ready to be punched.

An easy experiment can be performed to help understand how these steel blanks are annealed, and how annealing differs from hardening. The first step is to acquire an ordinary steel bobby pin, the kind your grandmother might have used to hold her hair in place. This bobby pin is then heated over the flame of a Bunsen burner or propane torch until it is red hot, and then quickly plunged into cold water. The steel will instantly harden. It will be very brittle and can easily be broken in half. If another steel bobby pin is heated and then allowed to cool slowly, it will behave differently. When it cools it will be much more malleable, bending without breaking. This steel has been annealed.

To understand the behavior of the bobby pin, it is necessary to discuss the two main allotropes or crystal forms of steel— body-centered cubic and face-centered cubic. In a body-centered cubic structure, each unit cell has an entire atom in the center of the crystal, and an atom at each corner of the cube, which is shared with adjacent unit cells. The unit cell of a face-centered cubic structure not only has a shared atom at each corner of the cube, but also has an atom in the center of each face, that is also shared with adjacent atoms (Figure 2.10). There is no atom in the center of the cube. The face-centered lattice dominates at higher temperatures. (As we saw earlier, gold also has this face-centered cubic structure.)

As the steel cools, it transforms into the body-centered cubic structure. As it has fewer atoms packed into the same amount of space, this structure is less dense, leaving room for carbon atoms to migrate in. Most steel alloys contain some carbon, which is added to increase hardness. Carbon atoms are a lot smaller than iron atoms, so a lot of carbon can be dissolved within the latticework of steel as they fill in the spaces between the individual unit cells. It is energetically favorable for the carbon atoms to occupy the space between the unit cells, but not the space *within* the unit cells. When the bobby pin is heated, the carbon atoms become much more energetic and move more freely, flitting about both between and within the individual unit cells. If the steel is cooled quickly by plunging into cold water these carbon atoms are locked into place wherever they might happen to be.

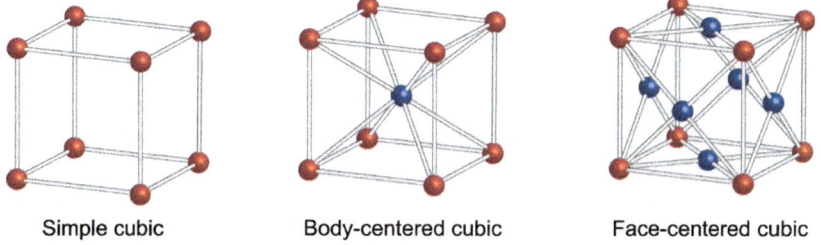

Simple cubic   Body-centered cubic   Face-centered cubic

Figure 2.10   Body-centered *versus* face-centered cubic structure.
© Shutterstock.

A number of them will be within the unit cells of the metal as well as between the unit cells. As these tiny carbon atoms are dispersed throughout the lattice of the metal, numerous irregularities are created within its crystalline structure, taking away its malleability and thus rendering it hard and brittle, making it easy to break.

To understand how this process works, consider a wall made from bricks and mortar—the bricks represent iron atoms and the mortar the carbon atoms. If the mortar could somehow migrate into the bricks themselves, the resulting structure would be more likely to be subject to breaking, and wouldn't have the "give" you might expect in a wall in which the mortar was in its proper place.

In metals that have been annealed, however, there is ample time for the carbon atoms to migrate out of the spaces within the crystalline structure, causing the resulting metal to be malleable, as there is minimal disruption to the crystal lattice.

When steel products are made, they are often subjected to an additional step, resulting in tempered steel. Tempering involves heating a metal until it is red hot, and then quenching it in cold water. It will then be heated again, albeit gently, to a lower temperature than to the temperature that it was originally heated, and then allowed to cool slowly. In this way, the right combination of hardness and malleability can be achieved to meet the needs of the product being made.

Once a coin blank has gone through the annealing process, it will again be covered with a surface layer of oxidation. Before coins can be struck this layer of surface oxidation must be removed. The annealed blanks are immersed in a pickling bath containing an acidic solution, such as dilute sulfuric acid, along with other substances that have an abrasive action. The Royal Mint uses ball bearings in the pickling solution to help remove the oxidation. The US Mint at one time used potassium bitartrate ($KC_4H_5O_6$), an acidic salt of tartaric acid (commonly known as cream of tartar), as part of the pickling solution. Its mild abrasive properties coupled with its low pH still make it a popular choice for novice coin enthusiasts who might want to clean their coins; it works best when used in conjunction with an acid such as vinegar. Today the US Mint uses a mild detergent to remove this layer. After going through the pickling process the shiny coin blanks are then dried.

2.16 STRIKING THE COINS

Before being struck, the blanks have to take a trip through the upsetting mill, in which each will receive a raised rim. The coins are dumped into a funnel-shaped container and pass at high speeds through a groove that becomes successively narrower as they pass through. A difference as low as 0.25 mm is sufficient to produce the raised rim. Owing to the high speeds at which they are propelled through the spinning grooves, they don't get stuck. The metal has nowhere to go but upwards, creating the neat raised rim, the defining feature of all modern coins. Once the blank receives the raised rim it is known as a planchet. The raised rim makes it easy to stack the coins; without the rim they would not stack nearly as well.

An easy way to illustrate the physics of this process is to use an adjustable metal hose clamp from the hardware store to cut out a cylinder of Play-Doh. The Play-Doh is placed in the center of the clamp, which when tightened gets smaller. The Play-Doh will be squeezed upward, just like a coin going through the upsetting mill.

If a coin is to be made from a single alloy of uniform composition then it is ready to be sent to the stamping press, from which it will emerge as a finished coin. (We will examine layered coins in the next chapter.) Hardened steel dies will impress an image onto both sides of the coin. Traditionally, the anvil die, which is stationary, creates an image on the reverse side of the coin, and the observe side is stamped with the movable hammer die, with both striking simultaneously. However, there is no rule that says which die must strike which surface, so it can work either way. At the US Mint in Philadelphia, the obverse image is struck by the anvil die and the hammer die creates the reverse image. Some mints have the dies mounted above and below the coin, while in others the dies are mounted horizontally. At the Royal Mint, 850 coins can be struck per minute, at an average force of 60 metric tons. If any reeding is to be done along the edge of the coin, the coin will be restrained within a collar that will impart the design to the edge. Coins that come off the stamping press are quite hot, as frictional forces convert kinetic energy into thermal energy each time the die strikes a coin. These coins have literally been tried by fire but have emerged shiny and flawless, each one in "mint" condition.

The alignment of the dies will determine the orientation of the coin when it is spun around. If a coin is rotated horizontally and the obverse and reverse images are both upright, then the coin has a medal die axis. If a medal is hung around the neck you want both sides to be upright when the medal is turned, hence the name. On the other hand, if a coin needs to be flipped vertically in order to view both the observe and reverse sides in an upright position, then the coin has a standard die axis, or coinage axis.

The earliest coins had little consistency when it came to die alignment, and it wasn't until the second century BC that most coins minted were deliberately aligned either one way or the other.[22] Most ancient coins had a coinage axis, but there were exceptions. The Greek tetradrachms were typically minted with a medal die axis. Examination of the die axis is one way to detect a counterfeit coin—if it differs from other coins of the same type it is almost certainly a fake. Most modern coins have a medal die axis, but there are still a number of countries that produce coins with a coinage axis. All American coins still display a coinage axis.

In examining a large container of coins from around the world, I discovered that those from Mexico, Brazil and the Dominican Republic all have a coinage axis. Before the euro, coins from Italy, France, Spain, Belgium and the Netherlands had a coinage axis as well. The euro has a medal die axis. Occasionally a few coins will enter circulation that have a different axis than what was intended. These mint errors are exceedingly rare, however, and are highly sought after by collectors. Therefore, flip through your change from time to time to see if you can find a coin that is oriented differently than the others—if you find one you may have struck it rich.

REFERENCES

1. K. M. Dunn, *Caveman Chemistry: 28 Projects from the Creation of Fire to the Production of Plastics*, Universal Publishers, 2003, p. 109.
2. C. J. Howgego, *Ancient History from Coins*, Routledge, London, 1995, pp. 26–27.
3. F. C. Thompson and A. K. Chatterjee, Ancient Greek plated coins, *Nature*, 1951, **168**(4265), 158.

4. O. Jarus, Athenian wealth: Millions of silver coins stored in Parthenon attic, *Live Science*, 2015, https://www.livescience.com/51353-silver-coins-stored-in-parthenon-attic.html.

5. L. Torrisi, A. Italiano and A. Torrisi, Ancient bronze coins from Mediterranean basin: LAMQS potentiality for lead isotopes comparative analysis with former mineral, *Appl. Surf. Sci.*, 2016, **387**, 529–538.

6. D. Bourgarit and F. Bauchau, The ancient brass cementation processes revisited by extensive experimental simulation, *JOM*, 2010, **62**(3), 27–33.

7. E. Caponetti, A. Francesco, D. C. Martino, M. L. Saladino, S. Ridolfi, G. Chirco, M. Berrettoni, P. Conti, N. Bruno and S. Tusa, First discovery of orichalcum ingots from the remains of a 6th century bc shipwreck near Gela (Sicily) seabed, *Mediterr. Archaeol. Archaeom.*, 2017, **17**, 2.

8. A. Deraisme, L. Beck, F. Pilon and J.-N. Barrandon, A study of the silvering process of the Gallo-Roman coins forged during the third century AD, *Archaeometry*, 2006, **48**, 3.

9. F. Reiff, M. Bartels, M. Gastel and H. Ortner, Investigation of contemporary gilded forgeries of ancient coins, *Fresenius' J. Anal. Chem.*, 2001, **371**(8), 1146–1153.

10. S. L. Niece, Depletion gilding from Third Millennium BC Ur, *Iraq*, 1995, **57**, 41–47.

11. S. Olsen, K. Sheedy, T. Knowles, V. Luzin, U. Garbe and F. Salvemini, Neutron texture and tomographic study of ancient Greek incuse coin production. 55-55. Abstract from Asia Oceania conference on neutron scattering (2nd: 2015), Sydney, NSW, 2015.

12. A. Pitarch and I. Queralt, Energy dispersive X-ray fluorescence analysis of ancient coins: The case of Greek silver drachmae from the Emporion site in Spain, *Nucl. Instrum. Methods Phys. Res., Sect. B*, 2010, **268**(10), 1682–1685.

13. G. F. Carter, E. R. Caley, J. H. Carlson, G. W. Carriveau, M. J. Hughes, K. Rengan and C. Segebade, Comparison of analyses of eight Roman orichalcum coin fragments by seven methods, *Archaeometry*, 1983, **25**(2), 201–213.

14. G. F. Carter and H. Razi, Chemical composition of copper-based coins of the Roman Republic, 217–31 B.C., in *Archaeological Chemistry IV*, ed. R. O. Allen, 1989, pp. 213–230.

15. C. Mancini and P. P. Serafin, Identification of ancient silver-plated coins by means of neutron absorption, *Archaeometry*, 1976, **18**(2), 214–217.

16. M. Baldassarri, G. H. Cavalcanti, M. Ferretti, A. Gorghinian, E. Grifoni, S. Legnaioli, G. Lorenzetti and V. Palleschi, X-Ray Fluorescence Analysis of XII–XIV Century Italian Gold Coins, *J. Archaeol.*, 2014, **2014**, 1–6.

17. S. Pepys, *The Diary of Samuel Pepys*, ed. R. G. Braybrooke and H. B. Wheatley, Complete 1666 N.S. Project Gutenberg, Release # 4200, 2004, pp. 921–925.

18. T. Levenson, *Newton and the Counterfeiter: The Unknown Detective Career of the World's Greatest Scientist*, Mariner Books, Boston, Mass, 2010, p. 87.

19. Z. Pelta-Heller, *Sir Isaac Newton & the Philosopher's Stone*, Chemical Heritage Foundation, Philadelphia, https://www.biography.com/news/isaac-newton-alchemy-philosophers-stone, 2017.

20. M. Keynes, Balancing Newton's mind: his singular behaviour and his madness 1692-93, *Notes Rec. R. Soc. London*, 2008, **62**(3), 289–300.

21. E. G. V. Newman, The gold metallurgy of Isaac Newton, *Gold Bull.*, 1975, **8**(3), 90–95.

22. L. R. Laing, *Coins and Archaeology*, Schocken Books, New York, 1970, p. 9.

CHAPTER 3

The Changing Face of Our Coins

"I believe in the gold standard.
I like solid lumps of things.
You can always melt them down."

Suzy Parker

Coins have come a long way since the first crude chunks of electrum were fashioned into money some 3000 years ago. Today, each coin is a flawless little masterpiece, a shiny work of art with a bit of history thrown in. However, coins are changing faster than ever, creating an ever-widening gap between the money of today and its forbearers. Although the methods used to make coins have maintained a remarkable degree of similarity, the materials themselves have undergone drastic changes. Of the original three coinage metals—gold, silver and copper—the only one still going strong is copper. Very few modern coins are made from gold or silver.

3.1 BYE, BYE GOLD

With the stroke of a pen, each nation would forever put an end to gold as the face of its currency, beginning with the UK, which abandoned the gold standard in 1931. Most other countries followed suit soon after, putting a stop to the production of gold coins as circulating currency. In the US President Franklin

The Chemistry of Money
By Brian Rohrig
© Brian Rohrig 2021
Published by the Royal Society of Chemistry, www.rsc.org

D. Roosevelt would issue an executive order making it illegal for private citizens to even own gold. This prohibition would stand until 1974.

It is only fitting that the last circulating gold coins ever made were among the most beautiful. The last US gold coin minted for general circulation was the "eagle," with a face value of $10.00. However, as is true for all gold coins, the actual value of the gold dwarfed the face value of the coin itself. A number of double eagles, $20.00 gold coins, were also produced in 1933, but as they were made after Roosevelt's executive order they were never put into circulation. All but a few were melted down, making the remaining specimens among the most valuable coins in the world, with one recently purchased by a collector for $7.59 million USD. The double eagle, designed by the sculptor Augustus Saint-Gaudens, is often considered to be the most beautiful US coin ever produced (Figure 3.1).

Gold coins are still produced today, but only as commemorative or bullion coins not meant for general circulation. Bullion coins are bought and sold on the commodity market. A bullion coin is issued by the government, has a designated face value, and is considered legal tender, but does not circulate. Its value is determined by its precious metal content. Any gold coin can act as bullion, regardless as to when it was minted, as long as the gold content is high enough. In the UK, the gold content of a coin must be 90% before it can be classified as bullion.

Perhaps the most famous gold bullion coin is the 1-ounce South African Krugerrand. At one point it accounted for 90% of the world's trade in gold bullion coins. The most popular British

Figure 3.1 Antique twenty dollar double eagle gold coin.
© Shutterstock.

bullion coin is the 1-ounce Brittania, with a face value of £100. In 1982 China issued a series of very popular Gold Panda bullion coins, with sizes ranging from 1-ounce down to 1/20 oz. The most popular gold bullion coin in the US is the American Eagle, with a face value of $50.

In 2007, the Royal Canadian Mint struck a 99.999% pure gold bullion coin with a mass of 100 kg. It had a diameter of 53 cm, and was 3 cm thick. Queen Elizabeth II is featured on the obverse side, and a maple adorns the reverse. Although it is legal tender, with a face value of $1 000 000, the value of the gold itself is much greater. Only five of these coins were made and offered for sale, which at the time were the largest coins in the world. When asked why they made such a large coin, Mint officials simply stated, "because we can". Unfortunately, the coin was not so large that it couldn't be stolen, as one of the coins was brazenly lifted, in a heist worthy of a movie script, from the Bode Museum in Berlin in 2017.

The record for largest gold coin didn't stand for long. It was broken in 2011, when the Perth Mint in Australia made a 1000 kg bullion coin measuring 80 cm in diameter, with a thickness of 13 cm! Even though it had a face value of $1 000 000, the gold was worth 50 times this much. Queen Elizabeth adorns one side and a kangaroo the other. This coin, owing to its enormous heft, would be significantly more difficult to steal (Figure 3.2).

Figure 3.2 World's largest gold coin produced by the Perth Mint and weighing in at 1000 kg.
© Adwo/Shutterstock.com.

3.2 BRASS

As gold is seldom used anymore, brass has emerged as a modern substitute. The color of brass is remarkably similar to that of gold. Although its composition can vary widely, an alloy must contain both copper and zinc to technically pass for brass. The £2 commemorative coins issued in 1986 to honor the Commonwealth Games were made of nickel-brass (76% Cu, 20% Zn, and 4% Ni), which resembles gold.

The golden-colored Sacagawea dollar, minted in 2000 in the US, has a copper core that is plated with manganese brass (77% Cu, 12% Zn, 7% Mn and 2% Ni). These coins are beautifully rendered, and though they look like gold, they have a distinctly different feel, being harder, less dense and somehow less charming.

Perhaps it's the ubiquity of brass that makes it less appealing than gold—it's seemingly everywhere. Its biggest claim to fame is its incredible resistance to corrosion. In 2013 a brass trumpet was recovered from the wreck of the USS Houston, which went down during World War II. After lying submerged in salt water for nearly 75 years the trumpet, while not playable, was still in surprisingly good shape. The amazing corrosion resistance of brass has made it useful for a wide variety of applications, being used in everything from plumbing fixtures to doorknobs to zippers.

If you receive a golden-colored coin in your change it will almost certainly be made from brass, which is perfectly suited for use in coins. It is easy to make your own brass coins. When I was a chemistry teacher one of my favorite labs involved making brass pennies. The procedure is quite simple, although perfecting the technique takes some practice. A shiny copper penny is allowed to soak in a solution of hot sodium hydroxide to which has been added some powdered zinc. In a short time, the penny will be completely coated with silvery-colored sodium zincate which forms when sodium hydroxide reacts with zinc, as described in the following reaction:

$$Zn_{(s)} + 2H_2O_{(l)} + 2NaOH_{(aq)} \rightarrow Na_2Zn(OH)_{4(aq)} + H_{2(g)}$$

The silvery penny is then held with a pair of tongs and passed quickly through the flame of a Bunsen burner a few times.

Heating reduces the Zn^{2+} ions in the sodium zincate into neutral zinc atoms, which diffuse into the copper, creating a shiny brass-coated penny, looking very much like gold.

On more than one occasion, after having had my students perform this experiment, my trash can burst into flames afterward. It would usually happen sometime during the following period, when the paper towels in my trash can would spontaneously combust, much to the delight of my students. As moist zinc dust will spontaneously ignite in air, the paper towels provided the perfect fuel source.[1]

3.3 BYE, BYE SILVER

Silver held out a little longer than gold, but not by much. The last British silver coins were minted in 1946, and these only had a silver content of 50%. The last silver coins in the US were made in 1964, which were 90% silver. Every other country in the world discontinued using silver in their circulating coins around this same time, owing to rising silver prices and the promise of alternative metals that were far cheaper. The loss of silver in coins represented the end of an era (a very long one). With the removal of this last precious metal from circulating coins, the intrinsic worth of all coins minted henceforth would be trifling. Silver coins were now an anomaly, to be hoarded rather than spent.

With the removal of silver and gold from coins, it became necessary to find other metals that could hold up to the rigors of circulation. Over time, four metals would emerge as viable substitutes—zinc, nickel, aluminum and stainless steel. However, none of these metals would ever be as noble as silver or gold. However, each one possessed something of value that would lead to their eventual adoption as reliable alternatives.

3.4 ZINC

Zinc has been known since ancient times, most notably as a constituent of brass alloys such as orichalcum. In the 13th century, India was engaged in large-scale zinc smelting operations, in which they were able to reduce zinc from its ore (often zinc carbonate) by heating it in a covered container. As zinc has such a low boiling point (907 °C), any attempts to smelt it would have proven fruitless, as the temperatures needed to smelt zinc are much higher than 907 °C. The zinc would have simply boiled away. By heating in a closed container, however, the zinc could be recovered as it condensed onto the cooler lid of the container. The Indians eventually would devise elaborate condenser systems to collect the zinc that boiled off. Recipes have since been found that reveal a wide assortment of reducing agents that were used to help isolate the zinc from its ore, including mustard, turmeric, natron and borax.

Europeans generally recognize the German chemist Andreas Sigismund Marggraf as the rediscoverer of zinc for the Western world, in 1746. By heating calamine and charcoal he was able to isolate pure zinc metal. Calamine is a pink powder consisting mostly of zinc oxide with 0.5% iron(III) oxide, the active ingredient in calamine lotion.

Zinc is relatively soft, coming in at 2.5 on Moh's hardness scale, the same as gold and silver. However, unlike other coinage metals, zinc is not that malleable at room temperature—in order to confer any degree of malleability, it must first be heated above 100 °C. Below 100 °C, it is somewhat brittle. Owing to its susceptibility to corrosion, few coins are made from pure zinc—it is better suited for use as an alloy or as an inner core where it is protected from oxidation.

3.5 NICKEL

In America, nickel burst on the coinage scene in 1865. The Civil War had exhausted the nation's resources. Silver and gold were being hoarded as people feared a monetary collapse, so a cheaper substitute for coin-making was sought. Nickel fit the bill. Like gold, silver and copper, nickel is very durable, offering excellent corrosion resistance. It is also malleable, allowing it to be easily molded into coins. Nickel is not hard to find, being the fifth most abundant element on earth; however, most of it is locked up in ores.

Niccolite, the ore from which nickel is often obtained, derives its name from German miners who thought that it contained copper, due to its reddish color. However, they were not able to extract any copper from it at all, which is not surprising as niccolite is nickel arsenide (NiAs). Being superstitious folk, they concluded that the devil, who was referred to as Old Nick, was responsible, so they named it kupfernickle, meaning "devil's copper".

Nickel is commonly found in meteorites. A large percentage of the earth's nickel is mined from the Sudbury region in Canada, which may have been deposited there by a meteor impact. Nickel, along with iron, is a constituent of the earth's core, contributing to the earth's magnetic field. Recent studies have suggested that even though nickel makes up only 20% of the core, the core would not produce the earth's magnetic field without it. Nickel today is of huge economic importance, being used to make everything from stainless steel to rechargeable nickel-cadmium batteries. Owing to its varied oxidation states, nickel compounds and their complexes can be used to impart vivid hues to objects. Green glass often contains nickel.

As the cost of nickel is increasing it is being used less and less in coins, and when it is used it is often as a plating. In 2018, nickel cost about twice as much as copper. If an American nickel were made from pure nickel, it would be worth almost exactly 5 cents, providing a rare example of a modern coin whose value is equivalent to the cost of its constituent metals.

An enduring numismatic mystery involves ancient nickel coins, which were only made for a short time in the second century BC in the ancient region of Bactria (home to the Bactrian camel) in what is now central Asia. They were composed of 25% nickel and 75% copper—the same formulation used in many coins today. It is quite likely the Bactrians obtained their nickel from the Chinese, who had produced nickel alloys as early as the 4th century BC, making a highly durable alloy of copper, nickel and zinc known as paktong. No other nickel coins would be minted for 2000 years.[2] These cupronickel coins are especially intriguing in light of the fact that nickel was only considered to be an actual element in 1751, when Swedish chemist Axel Fredrik Cronstedt isolated nickel from niccolite by heating it with charcoal.

3.6 ALUMINUM

The first circulating aluminum coins were not issued until 1907, in the country of British West Africa. They had a hole in the middle and were valued at 1/10 of a penny. There are no ancient aluminum coins, or, for that matter, any other ancient aluminum objects. As far as we know, aluminum did not exist as a pure metal before the 19th century. This fact helps explain why conspiracy theorists went nuts when an ancient piece of aluminum was allegedly uncovered in Aiud, Romania in 1974 during an excavation of mastodon bones. The wedge-chunked hunk wasn't tiny either—it was over 20 cm long and had a mass of over 2 kg. Subsequent analysis showed it to be an alloy composed of 89% aluminum.[3]

The "wedge of Aiud," as it has been called, was hailed by some as proof that we were visited by aliens in the past, who had evidently found a way to isolate aluminum long before we did. The more likely explanation is that it was a broken-off tooth from an aluminum excavator bucket. However, the find was so intriguing precisely because aluminum is such a "modern" metal, even though it is the third most abundant element in the earth's crust (after oxygen and silicon).

Alumina, or aluminum oxide (Al_2O_3), is among the various compounds with which aluminum is bound within the earth's crust. Corundum, one of the hardest minerals known (9 on Moh's scale) is made from aluminum oxide. Rubies and sapphires are varieties of corundum, the colors arising from impurities in the crystal—chromium imparts the red color to rubies while iron and titanium give sapphire its blue hue.

However, the majority of the world's aluminum comes from bauxite ore. Bauxite varies in composition, being primarily composed of various aluminum hydroxide and aluminum oxide compounds. A lump of bauxite is very easy to identify, having little circles of varying compositions cemented together, vaguely resembling frog's eggs.

It is aluminum's high reactivity that binds it up with oxygen to form such strong compounds, compounds that resisted being separated for a long, long time. It is difficult to assign credit to the first person who isolated aluminum, because it was really a process occurring over much of the 19th century, during which various scientists built upon the work of others.

A good place to begin is with the English chemist Sir Humphrey Davy, who pioneered the electrolytic separation of compounds into pure elements. By passing electricity through molten compounds, he was able to isolate such pure elements as sodium, potassium, barium, calcium, strontium, magnesium and boron. In 1809 he applied an electrical current to a mixture of alumina and iron, producing a small chunk of a hard, whitish substance that he first named alumium. Davy would go on to change the name to aluminum, and finally to aluminium. Americans tended toward the aluminum spelling, however, and in 1925 the ACS (American Chemical Society) made aluminum the official spelling. However, in 1990, IUPAC officially adopted the aluminium spelling, which Americans to this day ignore.

Building on the work of Davy was Danish chemist Hans Christian Oersted. In 1825 he combined a sample of aluminum chloride ($AlCl_3$) with a potassium amalgam (an alloy of mercury and potassium) to create an aluminum amalgam, as the potassium reacts with the chlorine. Heating this mixture under reduced pressure enabled the mercury to vaporize, leaving behind a mixture of aluminum and potassium chloride, which could be separated after they had cooled. The reaction is as follows:

$$AlCl_{3(s)} + 3K(Hg)_{(s,l)} \rightarrow Al_{(s)} + 3KCl_{(s)} + Hg_{(g)}$$

Oersted was able to isolate only a tiny amount of aluminum, not enough to be practical. In 1827 Friedrich Woehler would improve on this process by reacting potassium gas with aluminum chloride to isolate a purer form of aluminum. This reaction is as follows:

$$AlCl_{3(s)} + 3K_{(g)} \rightarrow Al_{(s)} + 3KCl_{(s)}$$

This mysterious new metal was not only very tough and durable, but also very light, with a density of only 2.7 $g\,mL^{-1}$. Throughout much of the 19th century aluminum was more expensive than gold. Napoleon III was especially enamored with this wonderful new metal, and would serve his most distinguished guests with aluminum utensils. The wealthy would adorn themselves with aluminum jewelry. The tip of the Washington Monument, completed in 1884, was bedecked with a six-pound pyramid of pure aluminum, testifying to its exotic nature.

Others would improve on processes to isolate aluminum, but the real breakthrough came in 1886 when Charles Hall, an inventor and chemist, sought to find a better way to refine aluminum. He began his experiments while still a student at Oberlin College in Ohio, USA. Working in a primitive laboratory he constructed in a woodshed behind his parents' house, he successfully discovered a method to separate aluminum from alumina, finally enabling the production of aluminum to be commercially viable. Hall would go on to fame and fortune, founding the Alcoa Corporation, which is today the world's biggest supplier of aluminum.

Hall's method involved dissolving alumina in molten cryolite or sodium hexafluoroaluminate (Na_3AlF_6) and then using electrolysis to reduce the aluminum from it, producing the pure metal. Amazingly, this same method was simultaneously discovered on the other side of the world by French chemist Paul Heroult, and as a result they were given mutual credit for the discovery, which has since become known as the Hall–Heroult process. Not only were Hall and Heroult born in the same year (1863), they also applied for a patent in the same year (1886), and died in the same year (1914) as well, only 8 days apart.

It may seem surprising that aluminum offers such superior corrosion resistance, as aluminum is a fairly active metal, readily giving up its electrons. However, it is precisely this affinity to react with oxidizers, notably oxygen, that keeps it protected. The aluminum reacts quickly with oxygen in the atmosphere to form a protective layer of aluminum oxide, preventing further oxidation, and thus corrosion of the metal. However, this aluminum oxide layer is only stable at a pH range of 4–9. If a piece of aluminum is dropped into a strong acid (or base) there will be a slight delay in reacting as the aluminum oxide layer is stripped off, enabling the underlying pure aluminum to then react.

Surprisingly, aluminum is rather soft, coming in at 2.75 on Moh's scale of hardness. Like gold and silver, it is most commonly alloyed with other metals to improve its hardness and durability. However, there are still coins that are made mostly of aluminum. As aluminum is rather inexpensive, these are normally low-denomination coins, such as the yen and the peso. Other examples include the Israeli 1-agora coins, which were minted for three decades beginning in 1960. About 35 countries currently issue aluminum coins, including Belize, Pakistan, South Korea, China, Croatia and Peru. Before the advent of the euro, many European countries used aluminum coins as well.

At today's prices, you can buy 5 g of aluminum for only a penny. Aluminum coins are easily identified by their silvery color and extremely light weight.

In 1974, the US mint made over a million aluminum pennies, as a possible substitute for copper. The design was ultimately rejected, partly because it would be difficult for an aluminum coin to show up on an X-ray in the event that one was swallowed, as aluminum behaves very much like soft tissue when X-rayed, their low density making them somewhat transparent to X-rays. Almost all of these pennies were destroyed, with one being donated to the Smithsonian Institution. Occasionally, an aluminum penny will turn up, as a few were given out to government officials as souvenirs. However, it is illegal to own one, and the US government recently forced a man to turn over an aluminum penny he had inherited from his father, a former mint official. He had planned on selling it at auction, where it was expected to bring upwards of $2 million. In some cases, aluminum is still more expensive than gold.

Owing to their light weight, the aluminum coins, such as the 1-yen, can be used for some interesting experiments. With a mass of 1 g, the 1-yen coin only weighs as much as a paper clip. If you carefully place one on the surface of a cup of water, it will not sink, as the surface tension of the water is sufficient to hold it up. It sinks down onto the surface of the water to an astonishing degree, but owing to the strong hydrogen-bonding between the polar water molecules, is able to stay supported on the surface.

3.7   STAINLESS STEEL

Iron has been known to man, since, well, the Iron Age, but its use as a metal in coins has been limited, because everyone knows that iron rusts, violating the basic tenet of using only corrosion-resistant metals in coin-making. A rusty coin is pretty much worthless. Even though iron has been alloyed with all sorts of elements throughout the ages to make it harder and more durable, these alloys still rusted to some extent. However, all that would change with the development of stainless steel, originally called rustless steel.

Harry Brearley, a metallurgist from Sheffield, England, is generally given credit for the discovery of stainless steel, applying for a patent for his specific brand of steel in 1915. Although

Brearley was not the first to develop a nonrusting steel alloy, he was able to market and produce his version more effectively than his competitors, receiving the lion's share of the credit for its invention. To say that Brearley was fanatical in his quest for the perfect steel is an understatement. He was so single-minded in his pursuits that in his autobiography he failed to mention his wife and son.

There are numerous varieties of stainless steel, but all contain at least 10% chromium. The chromium reacts with oxygen in the air to produce a thin layer of chromium oxide, sealing off the underlying metal from exposure to additional oxygen. Interestingly, this protection of the underlying metal by a layer of surface oxidation does not occur when steel rusts to form iron oxide. Iron oxide compounds have a lower density than iron, so when a metal rusts it expands, causing it to flake off in layers and thus degrade the metal.

One of the first stainless steel coins ever made was the 1-lira Italian coin, which was made from the alloy acmonital, an acronym formed from the first words of the Italian phrase *acciaio monetario italiano*, which translates into Italian monetary steel. It was first introduced in 1939 as a replacement for nickel, which was needed to manufacture armaments for the war. Albania quickly followed suit with the hefty 8.1-g Albanian 1-lek coin, also made from stainless steel.

Acmonital is characterized by its high chromium content (17.5–19%), thereby offering excellent corrosion resistance. Unlike some types of stainless steel, it is a ferritic alloy, being attracted to a magnet. It contains only a very small amount of carbon, making it soft and ductile, perfect for making coins.

3.8   THE 1943 RUSTING PENNY

When the US first experimented with making steel coins, they did not use stainless steel—unsurprisingly, the results were less than stellar. In 1943, at the height of World War II, copper was desperately needed to make munition shell casings. Therefore, the composition of the penny, which was previously almost all copper, was changed to zinc-coated steel. As the steel core of the penny was to be galvanized with zinc and would not be exposed to the elements, it was assumed that a low-grade steel would be sufficient. As zinc is more active than iron it would act as the sacrificial anode, protecting the steel core. (We will discuss these concepts in more detail in the next chapter.) However, the zinc was spread so thin that the edges of the penny were not covered, resulting in the subsequent rusting of the pennies.

These steel pennies would also get stuck in vending machines, which were designed to reject any magnetic metals. To this day, the 1943 penny is the only American coin ever minted that is attracted to a magnet. As steel has a lower density than copper the steel pennies are lighter, weighing in at 2.70 g. Not being satisfied with these rusting coins, in the ensuing years the US mint would attempt to round up and destroy as many of these steel pennies as they could find, although many still remain in circulation. They can usually be found in coin shops, where they can purchased for around a dollar each.

Before the US decided to go with steel for pennies, glass was seriously considered as a viable substitute for copper. A few glass prototypes were even made by Corning Glass Works in New York, which were then stamped by the Blue Ridge Glass Corporation in Tennessee, being struck while the glass was hot and soft to prevent shattering (Figure 3.3). These glass coins were made from standard soda-lime-silica glass, which is quite durable. Initial plans involved incorporation of uranium oxide in the

Figure 3.3 Blue Ridge glass coin.
Image courtesy of Numismatic Guaranty Corporation (NGC).

glass, which would have made them fluoresce under a blacklight, becoming the first anticounterfeiting measure to use ultraviolet light. According to one of the plant managers, the project was discontinued because uranium was needed for the Manhattan project.[4] If you have a desire to view fluorescent glass you can purchase some from any antique store, where it goes by the name of Depression glass, fluorescing a lovely greenish-yellow as a result of the uranium oxide.

Modern British pennies are not made with stainless steel either. In 1992 the composition of the penny changed from 97% copper to 94% steel. The newer pennies would contain just a thin copper coating; however, it was considerably thicker than the zinc coating on the 1943 US pennies (which only made up 1% of its mass). British pennies are made from mild steel, an alloy containing very small amounts of carbon, along with a little manganese. The low carbon content makes the metal more malleable and easier to strike; higher carbon steels would be much too hard. Even though mild steel has less tensile strength than higher carbon steel, making them less durable, they are nonetheless perfectly suited for making coins. Mild steel can rust, however, so it must be plated with another element that resists corrosion. After 2011, the composition of the five and ten pence coins changed as well, which would now be composed of nickel-plated steel. All of these steel coins are attracted to a magnet.

There are some modern coins made from stainless steel that do have an attractive silver color, however. Beginning in 1988, the Indian 50-paise coin, equal to half of a rupee, began to be made from ferritic stainless steel. Stainless steel coins have been minted by countries across the world, including Brazil, China, India, France, Iraq, Mexico, Ukraine, and Turkey.

3.9   THE BIRTH OF ELECTROPLATING

Throughout history, the majority of coins produced were alloys— homogeneous mixtures of metals uniformly distributed. About midway through the 20th century the landscape began to change, with coins of distinct layers becoming more commonplace. Nowadays it is more common to see a layered coin than a homogeneous one. There are several methods of producing these heterogeneous coins. If the outer layer of a coin is extremely thin it was most likely deposited on the surface of the planchet by some type of plating process. Although plating (and related methods of producing layered coins) are much more common today than in the past, by no means are multilayered coins a modern invention. As we have seen, there are any number of ancient Greek and Roman coins that were plated using a variety of techniques.

Plating today, though, typically refers to electroplating, which is the deposition of a thin layer of metal on the surface of an object using an electric current. While electricity is considered a somewhat modern phenomenon, its existence has been known throughout history. Lightning is of course a type of static electricity, which was well known before Benjamin Franklin flew his kite in a thunderstorm in 1752. The ancient Greeks were the first to rub fur on amber to produce a static charge.

If you visit the Baghdad museum you can find what looks like an ancient battery. In 1936 several clay pots were discovered near Baghdad that each contained a sheet of copper rolled into a cylinder, with an iron rod in the center. Speculation abounded that these urns were ancient galvanic cells. Even MythBusters (a popular American television show) got in on the act, building working replicas of these batteries. Using lemon juice as the electrolyte, they demonstrated that a single urn could generate 0.5 volts of electricity, and ten of them connected in series could produce 4.5 volts. However, having the potential to produce electricity and actually producing it are two very different things. The pieces of metal could have been put into the pots for any number of reasons. As there is no evidence of electroplating in the ancient world, if anyone did succeed in making a battery, there is no record as to what they might have done with it.

The first recorded instance of electroplating occurred in 1805, not coincidentally just a few short years after the first battery was made by Alessandro Volta, who gained inspiration for his device by studying the electric eel. He even named his battery *organe électrique artificial*, which is French for electric artificial organ, referring to the organ in an electric eel that generates current. His battery was a voltaic pile, consisting of alternating pieces of zinc and silver discs separated by cardboard soaked in salt water. As zinc and silver differ in how tightly they hold on to their electrons, if a conducting path is set up between them electrons will preferentially flow toward the metal that "wants" electrons more than the other, flowing away from the metal that doesn't want them as badly.

Luigi Brugnatelli, an Italian chemist, is generally given credit for being the first person to successfully electroplate an object (in 1805), although various others had experienced limited success before this. He described his findings as follows:

> I have lately gilt in a complete manner two large silver medals, by bringing them into communication by means of a steel wire, with a negative pole of a voltaic pile, and keeping them one after the other immersed in ammoniuret of gold newly made and well saturated.[5]

If you want to make your own "ammoniuret of gold," you are in luck, as Brugnatelli furnished the recipe:

> To one part of the saturated solution of gold in nitromuriatic acid, add six parts of solution of ammonia, by which the solution is decomposed and oxide of gold is precipitated, and a portion is set free, forming ammoniuret of gold. (Excerpt reproduced from ref. 6 with permission from Springer Nature, Copyright 1973.)

Brugnatelli experimental procedures were somewhat hard to follow, however, and replication proved difficult. However, by showing that electricity could be used to quickly and easily deposit a thin film of metal onto another metal, gilding never looked so easy. Or so profitable. It wasn't long before every would-be chemist with a battery and some wires was trying to perfect the process in order to make it economically viable. Among the first to succeed was George Wright of Birmingham, England, who discovered that potassium cyanide was a very effective way to dissolve silver and gold. In 1840 he collaborated with cousins Henry and George Elkington, who helped him to perfect the process. A 1908 account in *The Engineering Digest* describes the procedure they used:

> The Elkington solution was made by dissolving salts of gold in potassium cyanide; in their arrangement the articles to

be gilded form the cathode, a plate of gold the anode, both immersed in the cyanide bath. The gold as deposited on the article to be gilded was dissolved from the gold anode plate, thus keeping the auro-potassic cyanide bath of about the same strength; here we see that a solution of potassium cyanide, plus electricity, was at this early date (1840) a known and recognized commercial gold solvent, and with but slight modifications it is so used in electroplating today.[7]

The Elkingtons would go on to found Elkington & Co., one of the most successful electroplating companies in the world, specializing in silver plating. They would advertise their cutlery as being composed of a 35-micron thick layer of 99.999% "organically plated" silver.[8] Silver plating would become so successful that it would largely replace pure silver cutlery, as well as tea services, producing items just as durable and just as visually appealing for a fraction of the cost. They even supplied the plated tableware for the Titanic. Several forks and a butter knife would eventually be recovered from the wreckage of the Titanic, and besides being covered with a layer of green patina, revealing the presence of copper, were in surprisingly good shape.

It didn't take long for electroplating to catch on. In 1913, the American Electrochemical Society compiled a list of 193 different procedures for electroplating an object with gold alone,[9] and that was just a start—electroplating could work with virtually any metal. Plating was performed primarily for corrosion resistance, but was also performed to improve appearance and cut costs—even the thinnest of layers could provide all the benefits you might need from a metal, be it luster or conductivity. Today everything from hubcaps to jewelry is electroplated. The substrate does not even have to be another metal—even plastics can be electroplated. The plastic is first dipped into a metal bath to create a very thin metallic layer on its surface, with another metal being electrodeposited onto this layer.

Electroplating is not just for small objects—virtually any sized object can be electroplated. Perhaps one of the most ambitious plating projects ever undertaken was on the five bronze domes of the church of the Redeemer in Moscow, beginning in 1854. The largest of the domes was 30 meters in diameter. Even though it

Figure 3.4   The Dome of the Rock on the Temple Mount in Jerusalem, Israel.
© Shutterstock.

only required 28 g of gold per square meter, the project successfully plated almost 500 kg of gold on the surface of the
domes, taking 3 years to complete. The domes were not plated all
at once, however, but rather were taken apart and done in pieces,
using huge vats containing thousands of liters of cyanide
solution.[10]

The Dome of the Rock, the historic mosque in Jerusalem that
has stood since the 7th century, was refurbished in the early
1990s. The brass sheets that comprise the dome were electroplated with copper, then electroplated with nickel, and finally
electroplated with 80 kg of pure gold, spread out over a 3-
micrometer thick layer, making it the largest gold-covered
dome in the world (Figure 3.4).

3.10   ELECTROTYPES

It didn't take long for counterfeiters to discover that electroplating could be used to produce fake coins. As early as the 1850s
electrotypes of old coins began to appear. An electrotype is a copy
of an object made by the electrolytic deposition of a metal. It is
not hard to make an electrotype; the difficulty lies in making a
really good one. The first step is to make a wax impression of a
coin, which is then coated with a conducting medium of powdered metal or graphite. Molds are made of both the obverse and
reverse sides of the coin, and metal is electroplated onto the

molds. After being filled with another metal of suitable density, the two molds are then joined together. These fakes can sometimes be spotted by a joint in the middle if it is not filed down properly. Looking for this seam is one way to spot a cast coin. The next time you purchase a replica souvenir coin from a gift shop, look for the seam along the middle of the edge, a feature which will never be present on a genuine coin.

A properly executed electrotype can be quite convincing, as the electrodeposition of metal can capture very fine detail. Large copper coins are the most frequently forged, as copper is easy to fashion, being inexpensive and an excellent conductor. Probably the best way to detect a counterfeit coin is to check its mass and density and compare it to the published values of the coin in question. Electrotyping does have legitimate uses, however, often being used to copy printing plates. It has also been used to create replicas of metal sculptures, statues and medals.

Another electrochemical method used to create forgeries is known as spark erosion discharge. In industry it may be referred to as electrical discharge machining. It was first discovered by Joseph Priestly in 1770, who in observing the effect of sparks on metals noticed not only a corrosive effect but an erosive effect as well. At the time this phenomenon was a mere scientific curiosity, the usefulness of which would not be realized until a later date. The process begins with two metals that act as electrodes being placed a hair's breadth apart in a dielectric liquid. The dielectric liquid acts as an insulator between the two metal electrodes, allowing for a high voltage to be applied between them. Glycerin and deionized water are two common dielectric liquids that may be used.

The voltage between the two electrodes is then increased to the point at which the electrons in one of the electrodes eventually make the leap to the other electrode. It is this flow of electrons that creates a spark. A lightning bolt is a gigantic spark, which can of course inflict great damage. A little spark can also do damage to the surface of a metal, especially if these little sparks are striking at a rate of 100 000 times per second. The continual emission of sparks continues until the surface of one electrode has been negatively imprinted onto the surface of the other. This method is especially effective for etching fine details onto very hard metals, which may not be possible any other way.

A counterfeiter will use a genuine coin (or a quality replica) for one of the electrodes and a die for the other. The current is cranked up and the sparks fly. Raised areas of the coin will be closer to the die than recessed areas, creating a correspondingly stronger stream of sparks between these two areas, and thus a greater erosion of metal. A near perfect image of a coin can be engraved onto the surface of the die. This die is then used to strike counterfeits. Coins made from these dies are easy to spot—they will tend to have raised areas that are clearly visible under magnification, as the spark erosion process tends to produce tiny pits all over the negatively impressed die, especially if too much voltage was applied. To remove these raised areas, the coins need to be excessively polished, a telltale sign that something might be amiss, as the last thing one would want to do with a valuable coin is polish it.

Although modern coins are still sometimes counterfeited using electroplating technologies, a far more lucrative market exists in forging older coins. Some of today's best forgeries involve using sophisticated graphics programs to create an image of a coin, and then using a die engraving machine to make an actual die. In 2016 the American Numismatic Association reported a number of counterfeit gold coins that were flooding the market. Most seemed to have their origins in China.[11] Forged American Eagles and South African Krugerrands were showing up with alarming frequency, fetching huge prices. One common method of replicating these gold coins involved using a tungsten base and then electroplating it with gold. Tungsten was specifically chosen because its density is 19.3 $g\,mL^{-1}$, almost identical to that of gold.

3.11   THE 1982 US PENNY

The 1982 US 1-cent coin was one of the first plated coins to be circulated widely. Before 1982, the American penny was an alloy composed of 95% copper and 5% zinc and tin. Rising copper prices led to a shortage of pennies owing to people hoarding their cents, leading to corrective action by the US mint, which changed the composition of the penny in 1982. The core was now to be made mostly of zinc—99.2% zinc and 0.8% copper, to be exact—with just a thin copper coating, bringing the total copper composition to a mere 2.5% of the total mass of the coin. This new composition slashed costs. Pre-1982 pennies contained more than a penny's worth of copper, and as zinc was considerably cheaper than copper, the substitution made great economic sense. Best of all, the new pennies looked exactly like the old ones.

The layer of copper on these post-1982 one-cent coins is only 20 μm thick;[12] a layer this thin can only be deposited through electroplating. To illustrate the process, you can easily construct a simple electrolytic cell that will quickly and easily plate copper with zinc. The only materials required are a battery, two wires with attached alligator clips, a piece of zinc, a piece of copper and a solution of copper(II) sulfate. The zinc, acting as the cathode, is connected to the negative terminal of the battery and the copper, the anode, is connected to the positive terminal of the battery. The copper(II) sulfate solution acts as the electrolyte. Both the zinc and the copper, with attached wires, are immersed into the solution.

Reduction will quickly begin at the cathode, and oxidation at the anode. Reduction always involves a reduction of charge (hence the name), which is accomplished by the gaining of electrons. Oxidation always involves an increase in charge, which only happens when electrons are lost.

As electrons flow through the circuit, the zinc will acquire a negative charge as it is attached to the negative terminal of the battery. As a result, positively charged Cu^{2+} ions from the solution will be attracted to the negatively-charged zinc. Upon encountering these excess electrons in the zinc cathode, the positive copper ions are reduced into elemental copper, which will be plated onto the zinc. The reduction half-reaction is as follows:

$$Cu^{2+}_{(aq)} + 2e^- \rightarrow Cu_{(s)}$$

The longer the reaction is allowed to continue, and the stronger the current, the thicker the layer of copper that will be plated onto the surface of the zinc.

It is not difficult to remove the copper coating from a post-1982 penny. If you make some notches on the edge of a penny, you can expose its zinc core. If this penny is then dropped in hydrochloric acid, overnight the zinc core will be eaten away by the acid, leaving behind a paper-thin coating of copper, which will float, as it will be filled with hydrogen gas (at least initially). The reaction is as follows:

$$Zn_{(s)} + 2HCl_{(aq)} \rightarrow ZnCl_{2(aq)} + H_{2(g)}$$

The newer zinc pennies have created problems for young children, who sometimes put coins in their mouths and end up swallowing them. As hydrochloric acid in the stomach does not react with copper, if a child swallows a pre-1982 penny there is not much cause for alarm, and in the past doctors would usually advise anxious parents not to worry and just let nature take its course, as the penny would eventually pass. However, hydrochloric acid reacts vigorously with zinc. If the copper coating has worn thin on a post-1982 penny then the stomach acid can react with the zinc core, creating jagged edges that can potentially irritate the stomach lining or rupture the intestines.

I had a personal experience with this phenomenon when I was working at my desk in my home office one evening. My youngest child, who was about two years old at the time, was playing under my desk. I heard him gag, and then say, "Daddy, I ate a penny". I knew that he had swallowed a penny, and unfortunately, he did not observe the date before doing so. Having recently read about the dangers of swallowing these zinc pennies, we called our pediatrician, who said to bring him by in the morning.

She X-rayed his stomach, which revealed that the penny had already passed into the intestines. If it was still in the stomach, she said she would have tried to extract it (not a pleasant prospect), but since it had already passed, there was little she could do. The doctor said to keep an eye out when changing his diaper and to try to find the penny after it had passed.

That evening my wife recovered the penny while changing his diaper, approximately 24 hours after swallowing it. After cleansing it thoroughly we observed the date: 1984. Fortunately, it was still in good condition and there were no jagged edges despite it being an older zinc penny. It was oddly discolored, though, due to oxidation by powerful digestive juices.

There are many cases in the medical literature involving children swallowing pennies. In one case in 1998 at Duke

University Medical Center in the US, doctors treated a 2-year old who had swallowed a penny several days earlier and was now complaining of an upset stomach. X-rays revealed that the penny was full of holes, evidence that it had been eaten away by the stomach acid. The penny was removed by an endoscope. Subsequent weighing of the penny revealed that it had lost a quarter of its mass. The upset stomach was due to an ulcer in the stomach lining where the penny had been in contact.[13]

In another case in 2013, a three-year old toddler swallowed an elongated souvenir penny that was given out upon admission to the Bronx Zoo in New York. Two days later the penny was removed by a doctor, where it was discovered to have very sharp jagged edges, which had perforated the stomach lining. As the penny had already been flattened by a penny press, the copper coating was stretched thin, making it easier for the hydrochloric acid in the stomach to react with the zinc core within.[14]

Researchers in Vienna attempted to discover the toxic effects of swallowing the new euro coins. They soaked coins in simulated gastric juice for various intervals to see what might leach out. They tested 5 and 50-cent coins, as well as 1 and 2-euro coins, discovering the following:

> All coins underwent corrosive damage involving color alteration, erosion, margin chipping, and visible surface cavities. Their gross appearance immediately changed because chemical reactions with the simulated gastric acid induced bubble formation. Five-cent coins were especially prone to corrosion both in frequency and intensity. The weight loss measured for the various coins ranged from 0.10 to 0.41 g after 120 hours compared with baseline. All coins underwent corrosion… Holes were not observed. Metals contained in the coins went into solution after only 4 hours of exposure to simulated gastric juice. Depending on the type of coin, 2 to 6 mg of copper, 0.2 to 0.3 mg of nickel, 0.2 to 0.6 mg of zinc, 0.03 mg of iron, 0.2 mg of aluminum, and 0.02 mg of tin were dissolved. The concentrations of all dissolved metals increased progressively over time, but the greatest increase occurred within the first 48 hours. (Excerpt reproduced from ref. 15 with permission from Elsevier, Copyright 2007.)

Evidently, swallowing coins is nothing new. In the Middle Ages, if a foreign object became lodged in the esophagus, they had some very creative methods to remove it. One such method involved swallowing a small dry sponge on a string. Once the sponge expanded in the stomach, it would be removed, bringing with it (hopefully) the object that was swallowed.[16]

Dogs are especially susceptible to zinc poisoning from ingested pennies. As the zinc dissolves in the pet's stomach, it travels through the bloodstream, where it can cause anemia, as well as liver and kidney damage. If not treated, the dog can go into shock and die from organ failure. In 2013 a Jack Russell terrier puppy in Manhattan swallowed 111 pennies. Fortunately, he survived, as he was rushed to the vet where the pennies were promptly removed. Sadly, other cases have been reported where a single ingested penny has caused a dog's death.[17]

I once had my students bring in as many 1982 pennies as they could find, as I was curious as to how many were zinc and how many were copper, as both were made during this transition year. As they all contained the same date, we had to determine the composition of each by taking their mass. Of the 70 1982 pennies brought in, 50 of them had a mass of 3.1 grams, revealing that they were made of almost pure copper, and 20 of them had a mass of 2.5 grams, showing them to be the newer pennies made from zinc. The newer pennies have less mass as zinc has a density of 7.1 $g\,mL^{-1}$ compared to the density of copper which is 8.9 $g\,mL^{-1}$.

There are several other differences between the pre-1982 and post-1982 pennies.

Zinc has a much lower melting point than copper—zinc melts at 419.5 °C while copper melts at 1085 °C. If you hold a zinc penny with a pair of tongs and heat it over a Bunsen burner it will quickly melt, yielding a shimmery blob of molten zinc. If you heat a copper penny, the flame of the Bunsen burner will not be hot enough to melt it. It does turn a pretty shade of green though, due to excitation of the electrons from the copper ions.

Copper and zinc also sound differently when dropped on a hard surface, such as a granite countertop or a black lab table. Copper coins will produce a more high-pitched melodious ring with good resonance, and a zinc coin will produce a noticeably duller thud. Silver and gold coins will also produce a clearer ringing sound when dropped, similar to that of copper. The sound test is also a good way to detect counterfeits; to get around this, some makers of cast or electrotype forgeries will insert a thin layer of glass between the two halves of a coin to mimic this ringing sound when dropped or struck.

A very effective way to distinguish between a pre-1982 and a post-1982 penny is to place each in concentrated nitric acid. As toxic reddish-brown fumes of nitrogen dioxide are immediately produced that smell like chlorine, this demonstration is best performed under a fume hood or outdoors. A deep blue solution is produced as elemental copper is oxidized into copper(ɪɪ) ions. The pennies must be removed from the acid after a minute or so—if left in the acid for too long the coins will be completely gone. After being placed in the acid, the pre-1982 copper penny will be thinner, but the post-1982 penny will look almost black as the thin layer of copper quickly reacts, exposing the zinc core beneath. A little buffing will restore this zinc disc to a more lustrous silver color. The reaction of copper with nitric acid is as follows:

$$Cu_{(s)} + 4HNO_{3(aq)} \rightarrow Cu(NO_3)_{2(aq)} + 2NO_{2(g)} + 2H_2O_{(l)}$$

3.12   THE ELECTROPLATING PROCESS

The post-1982 US penny is not the only coin that has been electroplated, of course. Billions of electroplated coins are produced every year. As the planchets are plated before they are struck, it is not possible to tell a plated coin from one of a homogeneous composition by visual inspection. If electroplated, each plated layer will be of a single element. However, there is no limit as to how many different layers can be applied. The Royal Canadian Mint uses a multi-ply plating technique, plating alternating layers of nickel and copper over a steel core. In 2007 the Winnipeg facility alone struck over a billion multi-ply blanks, not just for Canada but for numerous other countries as well.[18]

The Royal Mint issues coins that are single, dual and multi-layered (3 layers).

In an attempt to learn more about the plating process, I made some inquiries with the Royal Canadian Mint, requesting some additional information. My requests were ignored. I had previously taken a tour of the Philadelphia Mint in the US and had hoped to speak with someone there, which didn't happen either. Next, I called the Perth Mint in Australia, hoping that the good blokes on the other side of the world would be more obliging. I was rebuffed with extreme prejudice. The lady I spoke with did have a pleasing accent, though, so it wasn't a total waste.

I did discover, however, that you can find out just about anything you need to know about a particular process by referring to a patent application, as they are all publicly available online. Even though the multi-ply plating technique of the Royal Canadian Mint is a patented process, they are not about to make it easy for you. The Royal Canadian Mint has filed 30 patents since 1990, with many of them containing overlapping information. Many of the patents detail multiple possible procedures, and many of these possible procedures themselves state that certain details are proprietary and therefore cannot be revealed. In addition, many of the procedures contain startlingly ambiguous language, making it impossible to know exactly what they are doing. Despite this deliberate attempt at obfuscation, there is still an immense amount of knowledge that can be gleaned from a patent application.

A review of patent #5139886 revealed a wealth of information about the details of one particular plating process. Before any electroplating can occur, the blanks must be cleaned thoroughly, as any corrosion whatsoever can severely impede the ability of the plating material to adhere. These annealing and pickling processes were described in detail:

> The blanks are then annealed at 700.degree.–900.degree. C. in an oxygen free atmosphere and cooled slowly... Without annealing it is found that the steel surface is oxidized easily upon pickling. The annealing under a hydrogen atmosphere helps remove the steel surface oxides.
>
> The blanks are then loaded into a rotary plating barrel. The number of blanks used in this development may vary from 90 to 200, depending on their sizes... Normal cleaning

practices prior to electroplating are used to prepare the blanks for plating. This may include any or all of the following steps: washing the blanks with special alkaline detergents, rinsing, solvent degreasing, electrolytic cleaning, and rinsing in deionized water.

The traditional method is to clean with a basic solution followed by an acid pickling which is supposed to improve the adhesion to the nickel or other coating. We have found it to be advantageous to reverse this procedure. We first use an acid pickling followed immediately by a quick wash with a dilute sodium hydroxide solution to buffer the acid. We have found that with the traditional cleaning procedure, there is some oxidation even if only a short time elapses before electroplating. We find that this oxidation is significantly decreased by reversing the order.

The pickling solution may be a 10% hydrochloric acid solution for 30 seconds at 55.degree. C. The solution is applied in the rotary barrel previously referred to, which is rotated at a rate of 10 rpm during cleaning pickling and rinsing. The rinse is with a mild basic wash to neutralize the acid. A suitable rinse solution contains sufficient sodium hydroxide to provide a solution of a pH of 9.0.

Next, the procedure to apply the first layer of nickel is given, which will make up 1% of the final mass of the coin, being 0.005 mm (5 μm) thick. The electroplating bath is composed of a mixture of nickel sulfate, nickel chloride and boric acid. "The blanks are flash coated with a nickel strike" for 30 minutes, at a current density of 8 ASF, which stands for amps per square foot, an abbreviation whose meaning I wouldn't have guessed in a million years—evidently the US is not the only country to have not completely converted to the metric system.

The next step is to apply the copper plating, which was to make up 6% of the total mass of the coin, at a thickness of about "20–30 μm". I was surprised that the figure was not more exact, but evidently an acceptable range is good enough. As with the nickel plating, an acid bath is preferred, which allows for a higher current density. The bath needed for the copper plating is mostly copper sulfate, along with copper metal, sulfuric acid and chloride ion (70 ppm). I had expected something more exotic than copper sulfate, which is readily available and dirt cheap, but if you're making billons of blanks every year you don't want something exotic. This step takes 4 hours.

These nickel–copper blanks are annealed again before the final layer of nickel is applied, at a thickness of 4–8 μm. The plating solution is similar, except nickel sulfamate is used instead of nickel sulfate. The final steps are as follows:

> Finally, the plated blanks are cleaned and minted or coined, that is to say, pressed in coining dies under impact force of the order of 170 000 to 200 000 p.s.i. to impart a suitable design to the surfaces and to shape the edges to provide a rim and sometimes a serrated edge. Next was the corrosion pit test, where they were immersed in a 2% sodium chloride solution for 4 hours. Again, most held up surprisingly well.

Next, the details of several tests are given, which are performed to make sure the coins could hold up under consumer use. One test is the humidity chamber test, in which coins are dipped in an artificial sweat solution for 72 hours. The majority of the coins held up well, with a note indicating that no rust had formed. Next a wear and tear test is performed for 8 hours, the details of which are not provided.

Among the conclusions offered near the end of the application was this:

> This practice promotes intermolecular bonding between dissimilar materials and through coating of the micropores. It also minimizes bridging and blistering as a consequence.[19]

Even US coins rely on the Royal Canadian Mint's expertise, as they have licensed their multi-ply plating technique to Jarden Zinc Products in Tennessee, the sole suppliers of the copper-plated blanks for US pennies. Only the striking of the coins is done at the US Mint. So as to ensure the consistency of the product, it is common practice for mints to rely on a single supplier for any specific application.

As of 2016, it cost 1.43 cents to make a one-cent coin in the US.[20] With every penny it strikes the mint is losing money, and it strikes a lot of them. In 2017 over 8 billion pennies were min-ted—58% of all coins made in the US. Every few years a move-ment to eliminate the penny gains some steam, but any attempts at doing so are invariably struck down, often owing to intense lobbying efforts by the metal industries.

Canada phased out its penny in 2012, as has Australia, Brazil, Finland, Norway and a host of other countries. The last time I was in Canada I noticed that upon receiving change it was always rounded to the nearest nickel, which I thought was rather con-venient. However, in the US the penny isn't likely to be going anywhere soon—it remains a cherished nuisance.

3.13 CLAD COINS

Perhaps the most popular coins today are clad coins, which are instantly recognizable by the appearance of different layers in cross section. With the phasing out of silver, clad coins came on the scene in a big way in 1965, with dimes and quarters being made with an inner core of pure copper and an outer layer of cupronickel (75% copper and 25% nickel) (Figure 3.5). If you look at the edge of an American quarter or nickel you can clearly see the alternating layers of copper and silver-colored alloy. As the

Figure 3.5 United States Washington Quarter.
© Shutterstock.

layers are visible, it could not have been made by a plating process, which will deposit metal uniformly along the entire surface of a coin. A clad coin is more like a metal sandwich. Clad coins are universally despised by coin collectors—they are seen as the ugly ducklings of coinage, strictly utilitarian with little aesthetic appeal, lacking the grace and elegance of the silver coins they replaced.

Immersion in nitric acid cannot only reveal whether a coin is plated or homogeneous, it can also reveal whether or not it is clad. If a US quarter or dime is placed in nitric acid, its copper core will quickly be revealed. If a homogeneous alloy is placed in nitric acid, however, such as the US nickel, it will get progressively smaller but will maintain the same look throughout.

The cupronickel alloy used in clad coins has a long and rich history, and has emerged as one of the toughest materials available for making coins. The first modern cupronickel coins were made in Belgium in 1860. The US followed suit shortly thereafter, minting three and five cent pieces made from cupronickel. As silver was scarce after the Civil War, its silver color made for a great substitute. However, cupronickel was known long before the 19th century. Some ancient Greek coins were made of cupronickel, as were some Bactrian coins.

Cupronickel is exceptionally resistant to corrosion, even in saltwater, making it especially useful in any type of application in which the metal will come into contact with seawater. On clad coins, the exposed copper along the edge seems not to matter much as far as contributing to corrosion; copper is a fairly noble metal in its own right, plus the surface area of the edge is small compared to its width.

The cupronickel alloys typically contain trace amounts of manganese. Manganese acts to make the alloy stronger, decreasing brittleness. It also serves as a deoxidizing agent, reacting with oxygen and sulfur to produce manganese oxide or sulfide. By removing these impurities and tying them up as harmless compounds, the overall strength of the alloy is increased.

Clad coins begin as three separate strips, which are joined together and then rolled to the desired thickness. Blanks are cut from this strip, and from there, they follow the same steps as any other coin. According to the US Mint, cladding is described as follows:

> Cladding bonds dissimilar metals together. The clad material could then be plated, but this is not usually done in coining. Typically, clad coinage metals are "roll clad," a process in which the layers of metal (for coins, this is usually a symmetrical "sandwich" of an odd number of layers) are thoroughly cleaned and passed through a series of rollers under sufficient mechanical pressure to bond the layers together. The high force serves to permanently deform the metals and to reduce their combined thickness in the process.[21]

Clad coinage in the US had a much more explosive beginning, however. From 1965–1971 the strips from which clad coins were made were supplied by the DuPont Corporation, who used a patented technique known as Detaclad. This technique involves joining metals together by a bizarre process known as explosion welding. Initial tests were carried out at the Los Alamos testing grounds, where the first atomic bomb was tested.[22]

A 1964 US patent filed by representatives from DuPont describes the process as follows:

> The present invention relates to a novel process for bonding of metals. Specifically, the invention involves a method for bonding metal surfaces and in particular for cladding a metal surface with one or more layers of the same or another metal...
>
> The use of clad, or composite, metals as materials of construction has become, in recent years, a well-established practice. Such clad materials consist of a base metal, usually relatively inexpensive, to the surface of which is bonded or clad a layer of a second metal which possesses certain desirable properties, *e.g.*, high corrosion or oxidation resistance, not characteristic of the base metal. In most instances, the metal which forms the cladding layer is considerably more costly than is the base metal to which it is applied. Hence a considerable economic saving is made

possible by the use of a thin layer rather than a thick layer of the costly metal.

A second advantageous feature of the use of clad metals results from the fact that frequently the metal possessing the desired corrosion resistance or other property is lacking in the necessary tensile strength, thermal properties, or compression strength to enable it to be deployed *per se* in applications where stress will be encountered.

The patent then describes the nature of the explosives used in this technology:

We have found that the above-described objects are achieved when we...place on the outer surface of said layer of cladding metal a layer of a detonating explosive having a velocity of detonation less than 120% of the velocity of sound in that metal in the system having the highest sonic velocity, and therefore initiate said explosive layer...

The metals must be separated from each other a distance at least sufficient for the explosively propelled layer to achieve an adequate velocity before impact with the stationary layer. A spacing of 0.001 inch between the facing surfaces of the two layers represents the minimum spacing which we have found will consistently be adequate...by increasing the explosive loading or by evacuating the space between the layers, spacings much greater than 0.001 inch are feasible. In general, however, separation of more than 0.5 inch is not convenient or necessary.[23]

Thirty-four separate examples are then presented, giving details of how to explosively join together various types of metals. Explosives include substances such as pentaerythritol tetranitrate (PETN) and TNT.

Explosion welding is a relatively new technology having only come on the scene in the 1950s. It was observed during World War I and II that shrapnel from shells that struck tanks at a high velocity would be permanently fused together with the metal plating. Explosion welding is especially effective in joining together metals that are incompatible and cannot be bonded together using conventional methods. It is also the most practical way to join together layers of metal that have large surface areas. Explosion welding has been used to make layered sheets of metal for spacecraft, railway locomotives, and a host of other heavy-duty vehicles and equipment. If an application involves high heat, pressure or the use of corrosive substances, such as you might find in an oil refinery or chemical plant, clad metal is often employed.

To begin the process, the layers of surface oxidation are removed and the two sheets are placed atop one another, separated by plastic spacers. A frame fits snugly on top which contains the explosives and will help to distribute the force evenly. Typically, a detonator is placed on one corner, and the explosion front travels diagonally across the surface of the metal, creating pressures of over 100 000 atmospheres (atm), permanently bonding the sheets of metal together. When the explosives detonate, a layer of gas is expelled from between the layers, driving out any oxides or other corrosion products, ensuring that the surfaces are free of any impurities before they are joined.

The metals at the point of contact are driven together with such force that they become plastic. (Plastic is the opposite of elastic—if something is elastic it snaps back to its original shape. When something is plastic it deforms permanently and does not snap back, like pulling on a string of taffy.) As a result, the delocalized sea of electrons that make up each metal atom's electron cloud become hopelessly enmeshed, forming metallic bonds that are often stronger than those in the parent metal.

Surprisingly, the explosion occurs so fast (detonation velocities are between 2000–6000 $m s^{-1}$) that little heat is transferred to the metal, even though significant heat is generated during the process.[24] As the metals do not experience a significant temperature increase they are not melted and experience little deformation, other than momentarily at the interface of the two metals during the explosion. Afterward, the clad metal is heated a little to soften it and then subjected to large presses or rollers to make sure the end product is uniform.

Explosion welding is actually a type of cold pressure welding. The first recorded observation of cold welding was by the Reverend J. T. Desaguliers in 1734, a British scientist and clergyman who once served as an assistant to Isaac Newton. He took two tiny lead spheres (25 mm each in diameter), and showed that by pressing them together and twisting they could be joined together.

The notion of putting pressure on two objects to facilitate bonding is actually rather commonplace. When applying a postage stamp or sealing an envelope flap I always pound on it to make sure I get good adhesion, regardless as to whether I am using an old-fashioned gum that needs licking or the new pressure sensitive seals.

Anytime two surfaces come into contact, there will be some type of intermolecular attraction. This same principle applies to friction—it is more difficult to slide a full bureau across a thick carpet than an empty one. The greater the weight of the bureau the more contact that exists between it and the underlying surface. Greater contact means greater van der Waals forces—those intermolecular forces of attraction that explain why water droplets adhere to windows and geckos can walk on ceilings—so, when we talk about greater friction acting between two objects, we are really saying that the intermolecular forces of attraction are greater. One could argue that just as friction keeps your ring from sliding off your finger, so too does it hold together two pieces of metal that have been squeezed together so tightly that there is no longer any space between them at all—the two objects have become one.

Not all changes in money are driven by the desire to deter counterfeiters, although historically that has been the number one driver of change, both for coins and for paper money. However, sometimes change is driven by other factors—most notably cost, which was certainly a factor when the US went from silver to clad coinage in 1965. There is another factor that must be considered anytime a mint is considering changing the composition of a coin, and that is whether or not the new coin will be accepted in vending machines.

3.14 VENDING MACHINES

The oldest record of a vending machine was one made by Hero of Alexandria, a Greek mathematician who was famous for his many inventions. Unlike other ancient Greek thinkers, he was not afraid to get his hands dirty, and embraced the experimental side of science. He invented all sorts of amazing machines, like steam engines, water pumps and mechanically operated puppets. A description of his vending machine, which dispensed holy water at Egyptian temples, is as follows:

> A person puts a coin in a slot at the top of a box. The coin hits a metal lever, like a balance beam. On the other end of the beam is a string tied to a plug that stops a container of liquid. As the beam tilts from the weight of the coin, the string lifts the plug and dispenses the desired drink until the coin drops off the beam.[25] (Jaffe, Eric. Copyright 2006 Smithsonian Institution. Reprinted with permission from Smithsonian Enterprises. All rights reserved. Reproduction in any medium is strictly prohibited without permission from Smithsonian Institution.)

Modern-day vending machines were introduced in London in the late 1800s, with the first one dispensing postcards. The first American vending machines appeared on railway platforms in New York, selling gum. From their outset, they had to be able to distinguish between real coins and counterfeit ones. Like Hero's machine, the early vending machines relied on a series of levers to discriminate between real and fake coins—if a coin was not heavy enough it wouldn't trip a lever. Slots were made of various sizes that corresponded to the coins of the day. The old-fashioned gumball machines utilized a series of gear mechanisms that only allowed coins of certain sizes to fit. All of these earlier machines were strictly mechanical in nature, relying on gravity and a panoply of simple machines to accomplish their tasks.

Today's electronic vending machines are a bit more complex. When you insert a coin in a vending machine, it falls downward and then rolls down a ramp about 10 cm long, depending on the machine. It is along this 10 cm journey that the machine will decide whether to accept or reject the coin. Even though there is a variety of vending machines, the technology they use to discriminate between real and counterfeit coins is similar. First, the coin passes through a set of optical sensors, which contain light emitting photodiodes situated on one side of the track and light receiving phototransistors situated on the opposite side. When a coin rolls down the track, this light beam is broken. The time that passes while the light beam is broken is used to calculate the size of the coin.

For a useful analogy, consider a solar eclipse, which a wide swath of the continental United States was fortunate enough to witness in 2017. From my location, 2 hours and 55 minutes elapsed between when the edge of the moon first became visible in the face of the sun until it passed. If the Earth had another moon that was smaller and the same distance away, it would not take as long to traverse this path. In the same way, a smaller diameter coin will block the light path for a shorter time.

The computer within the vending machine is actually measuring the secant of the coin at a predetermined height. (The secant is the segment formed by connecting two points on the outer edge of the circle, but not necessarily through the center.) By using some elementary physics formulas involving uniform acceleration and time, the secant of a coin can be established. If the secant of the coin measured does not fall within certain parameters of allowable coins, then the coin is rejected, falling through the rejection chute and returned back to you. Sometimes a perfectly good coin can be rejected—if that happens chances are it's not the diameter of the coin that is the problem, but rather that the coin failed the next test.

After passing through the light sensors, the coin then passes in between two electromagnets, which induce eddy currents in the coin. An analysis of the eddy currents will then be used to determine if the coin will be accepted. An eddy current is an induced electrical current that moves in small circles within a conductor, much like eddies that form in a stream. Eddy currents tend to be strongest on the surface of a conductor, only penetrating a short distance into the metal. A changing magnetic field is required to induce these eddy currents in a conductor, which can be accomplished in one of two ways—either by moving the magnetic field or moving an object through a magnetic field.

Just as a moving magnetic field gives rise to an electrical field, an electrical field gives rise to a magnetic field. Therefore, if a moving magnetic field induces eddy currents in a coin as it falls past the pair of electromagnets, a magnetic field will also be generated by the eddy currents in the coin. This induced magnetic field will slow down the coin as it falls. The stronger the eddy currents produced, the stronger the opposing magnetic field. The stronger the magnetic field, the slower the coin will drop.

The electromagnetic energy produced by these induced magnetic fields is propagated *via* electromagnetic waves. An analysis of the amplitude (height) of a wave can reveal information about the coin's conductivity. As every metal or alloy has a precisely defined conductivity value, if a coin that is deposited in a vending machine does not match the set of allowable values, it will be rejected.

The frequency of the electromagnetic wave can be used to determine a coin's magnetic permeability, or simply how readily it supports the creation of a magnetic field within itself. Ferromagnetic elements such as iron and nickel have high magnetic permeabilities, while those of insulators and diamagnetic metals are low. Every alloy has its own unique magnetic permeability. When the falling coin is subjected to the magnetic field of the electromagnets it falls through, the degree to which it becomes magnetized is compared to allowable values, and a coin is rejected if its magnetic permeability is not represented by one of these values.

More sophisticated vending machines will contain more than one pair of electromagnets, with each using a different frequency. Penetration depth changes with frequency, so it is important to have varying frequencies when dealing with coins of varying composition, such as bimetallic coins or coins composed of layers of different metals.

All of these factors together make up what is termed the "electromagnetic signature" of the coin, which is an easy way to refer to how the coin acts in a magnetic field. One patent application describes the electromagnetic signature as follows:

> Most coin acceptors presently in use rely on signals that result when a coin disturbs a variable electromagnetic field. For example, a coin moves between two coils acting as emitting and receiving antennae, respectively. The signal picked up by the receiving coil is then analyzed using a proprietary algorithm to produce what is called an electromagnetic signature (EMS) of the coin. Based on its EMS, the coin is either accepted or rejected.

A common problem affecting coin acceptors is the fact that electromagnetic signatures (EMSs) may be very similar for different coins. When the EMSs of coins of different denominations or coins issued in different jurisdictions are similar, there is an opportunity for fraud.

As referenced above, EMS values are not calculated by any physical, chemical or mathematical formula. Rather, they are a set of numbers generated by software and algorithms devised by each coin acceptor mechanism manufacturer. EMSs are unitless and are made up of a set of figures which are purported to determine the diameter, the edge thickness, the weight, the alloy composition, *etc.*, of a coin at different frequencies. Moreover, these values are not single repetitive values which identify the characteristics of the coin. Rather than being exact; the values vary from coin to coin within a certain range. Accordingly, that range is critical for coin acceptor manufacturers, since even perfectly valid coins may be rejected. The range of values must therefore be established so as to properly characterize the specific properties that identify the particular features of a coin, such as its diameter, edge thickness or alloy.[26]

Other sensors in vending machines may measure the acoustic signal a coin generates when it strikes a specific surface. As discussed previously, different metals sound different when dropped. Each coin has its own acoustic signature when dropped. This signature can be analyzed to determine if it matches the acoustic signature of allowable coins. Impact sensors are also used in some vending machines, which measures vibration and acceleration when a coin strikes a surface. Information about the elasticity of the coin can thus be determined, as a more rigid coin will strike with greater force than a coin made with a more forgiving metal. Measuring the impact of a coin can provide useful information about a coin's mass, as heavier coins will exert a greater force than lighter coins.

Vending machines are tailored to a specific country or region. In the United States, no circulating coins are attracted to a magnet (except for the 1943 steel penny) so any coin attracted to a magnet would be instantly rejected, as its magnetic permeability would be far too high. Close to the US-Canadian border, it is especially important to not put Canadian coins in American machines, and *vice versa*, as many Canadian coins are attracted to a magnet. When the 1943 steel pennies debuted, they would often jam vending machines.

Whenever coin composition changes, great care is taken to ensure that the diameter and thickness dimensions of the new coins are the same as the old, as any change in size would pose the biggest challenge to vending machines, requiring not just recalibration but retooling. In 2012 the vending machine industry in the US estimated it would cost anywhere from $700 million to $3.5 billion to upgrade machines to accept coins with a different electromagnetic signature, even if they were of similar size and weight.[27] The rejection of aluminum as a replacement for copper in the US penny was due, in part, to objections from the vending machine industry, who felt that aluminum coins were much too light to work properly in their machines.

Even minor changes in coins can create chaos within the vending machine world. In 2012, the Canadian government changed the makeup of their one and two-dollar coins, causing them both to be a little lighter. The mass of the 1-dollar coin dropped from 7.0 to 6.27 g, while that of the 2-dollar coin dropped from 7.30 to 6.92 g. In both cases, nickel was replaced with steel, and as steel has a lower density than nickel, the coins ended up being lighter. These small weight changes caused them both to be rejected by most vending machines, incurring costs of millions of dollars to the vending machine companies as they were forced to recalibrate their machines.

If someone deposits a fake coin in a vending machine, the owner of the machine might lose a dollar or two, but in trading companies that deal with gold bullion and other precious metals, fraud could push their losses into the millions. Sometimes even experts are hoodwinked, so investment in a simple electroconductivity device can potentially save these companies a fortune. The same technology utilized by vending machines is employed by these instruments. These handheld devices are placed directly on a coin, generating a magnetic field that induces eddy currents within the coin. By analyzing the characteristics of the eddy currents, the electrical conductivity of the metal can be determined. These instruments can also detect small cracks or other imperfections that might exist, which will disrupt the eddy current flow.

Earlier we discussed how the market was being flooded with counterfeit gold coins that were composed mostly of tungsten. An electrical conductivity tester could have easily revealed these fake coins. The electrical conductivity of gold is 45 MS m^{-1} while that of tungsten is a mere 18 MS. (The symbol "S" refers to siemens and is the SI unit of electrical conductance, which is equal to 1 amp per volt. It is the reciprocal to the ohm, so it can also be represented as $1\,\Omega^{-1}$. It used to be called the *mho*, which is *ohm* spelled backward.)

The device cannot tell you what type of metal you have, however, but only its conductivity value. Unfortunately, there are numerous copper alloys that have an electrical conductivity very similar to that of gold, which could thus "fool" the detector.[28] It is always wise to use an additional method along with electrical conductivity. A quick density determination would quickly reveal that a copper alloy is not gold.

X-ray fluorescence is another common method used to authenticate gold bullion and other precious metals. Conductivity testers that rely on eddy currents do not work on ferromagnetic metals, however, as the large magnetic field of these materials would overwhelm the much smaller magnetic field created by the eddy currents.

Eddy currents are easy to demonstrate. If a Japanese 1-yen coin, which is made from aluminum, is placed on a tabletop and a strong neodymium magnet is quickly placed on the coin's surface, it will momentarily rise up off the table before falling back down. As the moving magnet approaches the coin, eddy currents are created in the yen, which in turn generates a weak magnetic field, enabling the coin to be attracted to the magnet.

A more dramatic demonstration of eddy currents involves dropping a strong magnet through a copper pipe (the magnet should be close in size to the diameter of the pipe). The magnet will appear to be dropping in slow motion as its magnetic field is repelled by the magnetic field set up in the pipe as a result of the induced eddy currents.

Airport metal detectors rely on electromagnetic induction to detect metals. If you accidentally leave some metal coins in your pocket, the magnetic field generated by the coils within the metal detector will induce eddy currents in your coins, which in turn generate their own magnetic field, which in turn is detected by the detector. Airport detectors can spot any metal, but are only calibrated to go off when a certain threshold is reached, so the small amounts of metal in say, your zipper or ring will not set it off.

3.15 BIMETALLIC CURRENCY

Although both plated and clad coins are obviously bimetallic, or trimetallic, to the numismatist the term bimetallic has a completely different meaning, referring instead to coins made of two different types of metals that are both clearly visible. One of the first bimetallic coins was made in 1636—the English Rose farthing, worth a quarter of a penny (Figure. 3.6). It was mostly made of copper but with a brass wedge resembling a pie piece driven through it. These coins were virtually immune from counterfeiting, which was rampant at the time. From 1684 tin coins were made with a square copper plug in the center, again to deter counterfeiting (Figure 3.7). Americans first experimented with bimetallic currency as early as 1792. They would drive a small

Figure 3.6 Damaged example of rose farthing of Charles I. Note the absence of the brass wedge that would have been driven through it.
Reproduced from https://commons.wikimedia.org/wiki/File:TH_A36CF5,_Rose_farthing_(FindID_234190).jpg, under the terms of the CC BY-SA 2.0 license, https://creativecommons.org/licenses/by-sa/2.0/deed.en.

Figure 3.7   Tin farthing with copper plug of William and Mary.
© Shutterstock.

Figure 3.8   A 500 lire coin, 1989, reverse. First modern bimetallic coin.
© DeAgostini/Getty Images.

hole in the center of a copper blank and then insert a silver plug
before striking.

   The first modern bimetallic coin was produced by Italy in
1982—the 500 lire coin (Figure 3.8). The inner ring was made
from bronzital, a golden-colored alloy containing 92% copper,
6% aluminum and 2% nickel. Bronzital is an Italian word that
meshes together the word bronze along with the elemental
symbols for nickel (Ni) and aluminum (Al). This particular type
of alloy, as is true for all aluminum bronze alloys, is extremely
tough and corrosion resistant. The outer ring of the 500 lire coin
is made from the alloy acmonital, a stainless-steel alloy.

Another bimetallic coin was introduced in Canada in 1996, a two-dollar coin affectionately named the "toonie". The original composition of the two-dollar coin was an inner disc made from aluminum bronze (92% Cu, 6% Al, 2% Ni) and an outer ring made from pure nickel. To make these coins, a blank of pure nickel is first formed. Then a hole is punched out of the middle, creating the outer ring. The inner coin is made separately, with little indentations along its edges. Striking and joining occur in one fell swoop. When the two pieces are joined together and struck, the malleable nickel outer ring flows into these indentations, joining the two pieces together. Theoretically, this bimetallic locking mechanism should make it very difficult to separate the two parts. According to the Royal Canadian Mint the inner core of the coin can withstand up to 181 kilos of pressure, or about ten times the pressure the average human hand can exert.[29]

Despite this ingenious locking mechanism, some of the earlier toonies could be separated if thrown against a hard surface. Necklaces would be made from the separated components. The Canadian authorities were not amused by these attempts, and passed laws making it a crime to separate the two parts of a toonie. Despite reports that they could be separated by freezing, I have found that even immersing a toonie in liquid nitrogen will not separate the components. However, I have been successful in separating them by heating a coin with a Bunsen burner and then plunging it in a beaker of water. It normally takes several cycles of heating and quenching before the inner piece pops out. Heating is able to separate the two components because different metals expand at different rates. A common physics demonstration involves taking a bimetallic strip consisting of a thin strip of nickel (or steel) on one side and bronze on the other, and then heating over a flame. The bimetallic strip will curl as one metal expands at a different rate than the other. The rate at which a metal expands when heated is determined by its coefficient of linear expansion. The metal that expands the most will be on the outside of the strip. (This technology was once commonplace in thermostats used in homes and in automobiles, but has since largely been supplanted by digital mechanisms.)

When the composition of the toonie changed in 2012, the inner core was still made from aluminum bronze, but now had a brass plating; the outer ring was changed to steel, with a nickel plating. Both versions of the coin look similar.

3.16 MAGNETIC COINS

A unique feature of the toonie, in both the original and current versions, is its attraction to a magnet. Until recently, virtually no coins were ferromagnetic. There are just three ferromagnetic elements at room temperature—iron, nickel and cobalt. Any ordinary magnet will be attracted to a ferromagnetic substance. If you place a magnet in a pile of coins, whatever sticks will be made from one of these three elements. If an alloy contains a majority of one of these metals, it will typically be attracted, but not always.

A bus driver in Alberta, Canada took advantage of the ferromagnetism of Canadian coinage by attaching a magnet to a long pole and removing coins daily from the coin collection unit on the bus, managing to collect over \$2 million dollars before he was caught.

Another magnetic Canadian coin is the aptly named nickel, which was made from pure nickel up until 1981, when the recipe was changed to that of the American nickel. American nickels have always been made from the cupronickel alloy; as this alloy only contains 25% nickel, these coins are not attracted to a magnet. In order to be ferromagnetic at room temperature, the nickel concentration must be at least 68.5%. In 1999 Canada changed the composition to steel, so their nickels are once again ferromagnetic.

Pre-1982 Canadian nickels provide an easy way to demonstrate when the Curie point has been reached, which is the temperature at which a substance loses its ferromagnetic properties. At 358 °C, nickel has the lowest Curie point of the ferromagnetic elements (the Curie point of Fe is 770 °C and that of Co is 1130 °C). If a magnet is grasped with a pair of tongs and attached to one of these nickels, it will quickly fall off when heated over a flame, as the Curie point will have quickly been reached. As soon as it cools though, it will regain its ferromagnetic properties.

To understand why heating destroys a metal's ferromagnetism, it is necessary to examine an atom's electron arrangement, specifically its 3d orbitals, which are the outermost regions within the third energy level where electrons are likely to be found. By examining how the electrons in these orbitals fill, the pattern quickly becomes evident (Figure 3.9). Electron spin can be represented by an arrow, which we can call "up" or "down" and can be represented by either a ↑ or ↓. By convention, arrows pointing up refer to a clockwise spin.

In each case in Figure 3.9, there are unpaired "d" electrons, in which the electrons within these orbitals are only spinning in one direction. An electron spinning within an orbital in just one direction gives rise to a magnetic field. Another electron within the same orbital will always spin in the opposite direction, canceling out the magnetic field.

This principle can be vividly demonstrated with an ordinary flashlight battery, an iron nail, two pieces of insulated wire and a paperclip. If the wire is tightly wrapped around the entire length of the nail, and then its opposite ends attached to the terminals of the battery, a miniature electromagnetic is created, which will pick up the paperclip. The flow of electrons through the wire, all in the same direction, give rise to the magnetic field.

However, if another wire is carefully wrapped in the opposite direction around the first wire, with the exact same number of coils, then the magnetic field will be canceled out, and the nail will not pick up the paperclip. This set-up represents what occurs in a diamagnetic element, in which every d orbital is filled. If one examines the 3d orbital filling diagram for a diamagnetic element, such as zinc, it will look like Figure 3.10.

Each orbital contains a pair of electrons, each spinning in the opposite direction, canceling out the magnetic field, and thus rendering the atom diamagnetic, which is not attracted to a magnet.

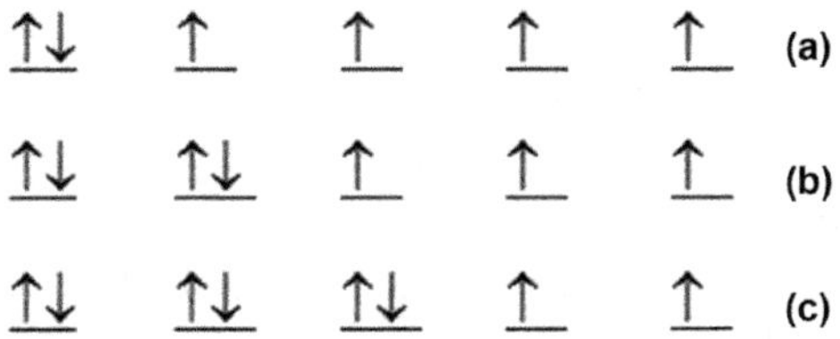

Figure 3.9 The 3d orbital diagrams for iron (a), cobalt (b) and nickel (c).

$$\uparrow\downarrow \quad \uparrow\downarrow \quad \uparrow\downarrow \quad \uparrow\downarrow \quad \uparrow\downarrow$$

Figure 3.10   The 3d orbital of a diamagnetic element.

When a large number of atoms align with the same orientation a domain is created. If sufficient domains align in the same orientation an observable magnetic field is created. When heated, the atoms move around rapidly and the domains become unaligned, causing the material to lose its ferromagnetism. When cooled, the domains rapidly realign and ferromagnetism is restored.

3.17   COLORED COINS

Among the many new innovations in coin-making is that of adding color not imparted by the metals themselves. In 1993, both Uganda and Equatorial Guinea minted the first colored coins. In order to be considered for this honor, the country issuing the coin must be a recognized nation and the coin must be legal tender. The color must be added by the mint itself or an agency authorized by the mint to apply the color. Therefore, if you or I were to simply paint a coin, it would not qualify as a true colored coin.

The Equatorial Guinea coin has a face value of 7000 francs, and features an orange dinosaur (Styracosaurus) against a multicolored background. The 2000-shilling Ugandan coin features the Matterhorn, and contains splashes of color with the purple mountain looming in the background. In the 1990s Papua New Guinea and Cuba followed suit with some beautiful colored coins. In 2004 Canada issued a quarter with a bright red poppy placed in the center of its reverse side. These were the first colored coins to undergo general circulation.

Coloring metals can be a tricky process. It is important that the ink does not wear off, and ideally should last as long as the life of the coin. The Perth Mint in Australia has made some beautiful colored coins, incorporating an innovative pad-printing technique. Their process begins with the making of four separate printing plates, each containing a different aspect of the design. Silicone pads absorb the ink from the plates and deposit it onto the surface of the coin in layers, with just the right mix of the four pigments until the desired color is produced. A hardener is then added to the ink on the coin, making the resulting image hard, durable and shiny.

Like any CMYK printing process, they begin with cyan, magenta, yellow and black. This process is subtractive in nature—colors are added that absorb (or "subtract") certain wavelengths of light. If all three colors—cyan, magenta and yellow—are added together, the result is black, which represents the absence of light, as all colors are absorbed. In actuality, though, these three colors produce more of a grayish hue, therefore if true blackness is the goal, some additional black ink must be added.

During CMYK printing, the beginning substrate is white or light-colored and different colors are added that absorb certain wavelengths of light. Each pigment will absorb all the wavelengths of its complementary color. As cyan will absorb red light, magenta will absorb green, and yellow will absorb blue, adding together all three will allow no light to escape, as all three primary light colors will have been absorbed. By filtering out certain wavelengths any color can be produced using only cyan, magenta and yellow. If you want red to be your end color, you would not use any cyan, as red is the complement of cyan—cyan filters out all red. However, if you were to mix magenta (which filters out green) and yellow (which filters out blue) then the only color left will be red.

Australia's beautifully rendered colored coins have been a big hit with just about everyone who has seen them. Everyone, that is, except the Royal Canadian Mint, who claimed that the Australian Royal Mint used their patented printing method without permission, specifically to make the $2 Remembrance Day commemorative coin featuring a bright red poppy. After inspecting these coins, Canadian officials concluded that the process the Australians used was essentially the same as the one they used to create a red poppy on their 2015 25¢ coin. The Canadian authorities filed suit, demanding that Australia either destroy all the remaining red poppy coins not in circulation or hand them over. In light of this lawsuit, the refusal of the Perth Mint representative to speak with me made more sense.

3.18   THE EURO

With the advent of the new millennium, diversity in the world's coinage took a big hit. The currency of some two dozen countries was replaced by a single currency—the euro. Although the euro was adopted in 1999, the first euro coins were not minted until 2002. Next to the American dollar the euro is the most popular currency in the world. Each denomination of coin has a common design on the reverse, with individual countries deciding what to put on the obverse. There are eight denominations—€0.01, €0.02, €0.05, €0.10, €0.20, €0.50, €1.00, and €2.00 (Figure 3.11).

The three lower denomination coins all have a similar design and are all bronze-colored, with a successively increasing diameter. They each have a smooth edge, with the €0.02 coin having a groove along its edge. Being made from copper-plated steel, they are attracted to a magnet.

Likewise, the €0.10, €0.20, and €0.50 coins look very similar to one another but different from the others. They are golden colored, being composed of an alloy known as Nordic gold, which contains no actual gold. Nordic gold made its debut in 1991 in Sweden with the 10-kronor, being composed of 89% copper, 5% aluminum, 5% zinc, and 1% tin. It is very tough and virtually corrosion resistant, retaining its golden color with virtually no discoloration. It is also hypoallergenic, transmitting few germs.

Figure 3.11   Complete set of Euro coins.
© Shutterstock.

The €1.00 and €2.00 coins look different yet, and are bi-metallic, with the inner part of each made from a sandwich of three metals. The core of the €1.00 contains an inner layer of nickel sandwiched between two layers of cupronickel. The core of the €2.00 is golden colored, and is made from a layer of nickel sandwiched between two layers of a nickel-brass alloy. The same material that makes up the outer silvery-colored layer of the core of the €1.00 coin makes up the outer ring of the €2.00 coin (cupronickel). Likewise, the same material that makes up the golden-colored outer layer of the core of the €2.00 coin makes up the outer ring of the €1.00 coin (nickel-brass).

Each of the four major coin types can be found within these eight denominations of the euro. The Nordic gold coins are made from a single homogeneous alloy. The lower denominations utilize electroplating to deposit the thin layer of copper onto the steel core. The more valuable coins are bimetallic, which are considerably more difficult to counterfeit. The inner core of these coins is made of clad metal. The unique patterns of reeding are designed to help the visually impaired, and the color differences make it less likely for a consumer to confuse a high-denomination coin with a lower one.

Just as the euro unified the currency in part of the European Union, the euro also represents a unification of sorts of the four major types of coins, as plated, clad, homogeneous and bimetallic varieties are all embraced. Their inclusion nicely shows that each type still has its place. Even though euros don't contain any gold or silver, their range of colors harkens back to a time when the three coinage metals of gold, silver and copper still reigned.

REFERENCES

1. J. Moszczynski, Sussex county school caused by accidental chemical mix, officials confirm, *The Star-Ledger*, 2012, https://www.nj.com/news/index.ssf/2012/04/sussex_county_school_fire_caus.html.
2. C. M. Schwitter and C. F. Cheng, Bactrian nickel and Chinese bamboo, *Am. J. Archaeol.*, 1962, **66**(1), 87.
3. S. Wagner, The most puzzling ancient artifacts, 2018, https://www.thoughtco.com/the-most-puzzling-ancient-artifacts-4086398.

4. C. Larson, The curious history of glass coins. Corning museum of glass, 2007, https://blog.cmog.org/2017/12/15/curious-history-glass-coins/.

5. E. Houston, *Electricity in Every-Day Life*, Nineteenth Century Collections Online (NCCO): Science, Technology, and Medicine: 1780–1925, P.F. Collier & Son, New York, 1905.

6. L. B. Hunt, The early history of gold plating: A tangled tale of disputed priorities, *Gold Bull.*, 1973, **6**(1), 16–27.

7. H. Frost and C. M. C. Sames, *The Engineering Digest*, H. Frost, New York, vol. 3, 1907, p. 152.

8. The history of Elkington & Co. & James Dixon, http://www.steelcitycutlery.com/aboutelkington.html.

9. J. C. Garcia and T. D. Burleigh, The beginnings of gold electroplating, *Electrochem. Soc. Interface*, 2013, **22**, 2.

10. L. B. Hunt, The early history of gold plating: A tangled tale of disputed priorities, *Gold Bull.*, 1973, **6**(1), 16–27.

11. H. Weisbaum, *Glitters, but Not Gold: Fake Gold and Silver Coins 'flooding' Market*, NBC News, https://www.nbcnews.com/business/business-news/glitters-not-gold-fake-gold-silver-coins-flooding-market-n591201, 2016.

12. E. Dickinson, Electroplating: How the U.S. Mint makes a penny, 2014, https://www.comsol.com/blogs/electroplating-u-s-mint-makes-penny/.

13. K. S. Pennies, Children's health issue, brief article, *Environ. Health Perspect.*, 1999, **107**, 6.

14. K. Boniello, Zoo sued after child swallows souvenir penny, New York Post, 2014, https://nypost.com/2014/02/02/bronx-zoo-sued-after-child-swallows-free-souvenir-penny/.

15. W. Rebhandl, A. Milassin, L. Brunner, I. Steffan, T. Benkö, M. Hörmann and J. Burtscher, In vitro study of ingested coins: Leave them or retrieve them?, *J. Pediatr. Surg.*, 2007, **42**(10), 1729–1734.

16. J. H. Chang and J. D. Burrington, Removal of coins from the esophagus: Nothing new under the sun, *Pediatrics*, 1973, **51**, 2.

17. K. Becker, Eating just one of these can be fatal to your pet, 2016, https://healthypets.mercola.com/sites/healthypets/archive/2016/07/30/pet-zinc-toxicity.aspx.

18. Royal Canadian Mint, Royal Canadian Mint credits patented multi-ply plating for record year in minting international

circulation coins, 2007, https://www.mint.ca/store/news/royal-canadian-mint-credits-patented-multiply-plating-for-record-year-in-minting-international-circulation-coins-5000068?cat=News+releases&nId=&parentnId=&nodeGroup=#.XAwXvS2ZPUo.

19. H. Truong and M. Dilay, Royal Canadian Mint, Patent #5139886, Coins coated with nickel, copper and nickel, 1991, https://patents.justia.com/patent/5139886.

20. M. Z. Donahue, How much does it really cost (the planet) to make a penny?, 2016, https://www.smithsonianmag.com/science-nature/penny-environmental-disaster-180959032/.

21. Alternative metals study, phase II, Technical report, 2014 biennial report to the Congress, United States Mint, Department of the Treasury, 29, 2014, https://www.usmint.gov/wordpress/wp-content/uploads/2016/06/2014-rd-biennial-report-appendix-4.pdf.

22. W. T. Gibbs, Daughter preserves father's record of 1964 coinage testing, 2014, https://www.coinworld.com/news/us-coins/2013/06/daughter-preserves-fathers-record-of-1964-coi.all.html.

23. G. R. Cowan, J. J. Douglass and A. H. Holtzman, Explosive bonding, *U. S. Pat.*, 3,137,937, 1964, http://www.freepatentsonline.com/3137937.pdf.

24. F. Findik, Recent developments in explosive welding, *Mater. Des.*, 2011, **32**(3), 1081–1093.

25. E. Jaffe, Old world, high tech. Smithsonian magazine, 2006, https://www.smithsonianmag.com/science-nature/old-world-high-tech-141284744/?no-ist=&page=2.

26. H. C. Truong, Royal Canadian Mint, Patent #9447514. Control of electromagnetic signals of coins through multiply plating technology, 2016, https://patents.justia.com/assignee/royal-canadian-mint.

27. C. Balakgie, Letter to U.S. Mint on behalf of NAMA; RE: Coin modernization, oversight, and continuity act of 2010 federal register request for comments, 2014, https://www.namanow.org/wp-content/uploads/NAMA-COMMENTS-to-the-U.S.-Mint-on-Metallic-Content-of-Coins.pdf.

28. S. Zaum, Authentication of gold bullion and coins. Talk given at the 10th Technical Forum, 43rd World Money Fair, Berlin, Germany, 2014, https://coinweek.com/featured-news/world-mints-10th-technical-forum-presentations-berlin-germany/.
29. Royal Canadian Mint, Balance and composition – the 2-dollar coin, 2018, https://www.mint.*ca*/store/mint/about-the-mint/2-dollars-5300016#.XA1BZS2ZN-U.

CHAPTER 4

Corrosion (and Related Matters)

"As iron is eaten away by rust,
so the envious are consumed
by their own passion."

Antisthenes

When a coin rolls out of the mint it is in pristine condition—new, shiny and uncirculated. It is in "mint" condition, but it won't stay that way for long. From the moment of its creation, every coin begins a sad and steady decline. This process of entropy accelerates when the coin enters circulation. Coins are literally wearing away—given enough time they will eventually wear away into nothing. The raised areas—the relief—begin to go first. The points that are raised the highest will erode the fastest. Eventually even the date will be unreadable. Then corrosion sets in, further degrading the coin. Banks no longer want these worn and mutilated coins so they sell them back to the mint. Returning home from duty, instead of receiving a hero's welcome these battle-worn coins are cast into a fiery furnace, cruel recompense for their many years of faithful service. However, even the most dirty and disfigured coin will find purification in the cleansing fire, and will one day be born anew as a brand-new coin, ready to begin the cycle all over again.

The Chemistry of Money
By Brian Rohrig
© Brian Rohrig 2021
Published by the Royal Society of Chemistry, www.rsc.org

4.1 PHYSICAL WEAR

According to the US Mint, the average lifespan of a circulating coin is only 25 years. However, Mint officials have little nostalgia for the past. Their only goal is to keep the money supply in tip-top shape. If not for coin collectors, fewer older coins would even be around. But truth be told, a lot of these older coins are in pretty bad shape. Those that are the least susceptible to corrosion are sadly the most affected by erosion. Despite millennia of efforts to make coins from the most durable alloys, even the best materials are no match for the rigors that a coin must endure. A coin made from harder metal is less likely to wear away, but if the metal is too hard it can't be used to make coins—hardened metals are simply too difficult to work with. Coins must be made from metals malleable enough to be rolled but soft enough to be struck, making every coin in circulation vulnerable to mechanical wear and tear.

The biggest culprit is friction. Coins are handled constantly. They are dropped, stepped on, and put through the wash. They rub against other coins and keys in your pocket or purse, and against other metals as they fall through vending machines and parking meters. Considering all they have to endure, they typically fare pretty well.

To verify whether the assertion of the US Mint of 25 years for a coin's tenure was accurate, I tabulated all the dates from the big jar of coins in our kitchen—a combination of mostly pennies, nickels, dimes and quarters. I made two piles: those 25 years and younger, and those older than 25 years. Of the 278 coins, 212 (76%) were less than 25 years old, dating from 1993 or above. The oldest coin was a 1941 nickel. It was still in good shape, with surprisingly little wear. While a few of the older coins, especially the pennies, had significant amounts of corrosion, overall the appearance of the older coins was not much different from that of the newer ones, which makes sense if one considers that coins are not replaced owing to age, but rather due to condition. It is quite possible that many of these older coins were perhaps lying in a drawer for a number of years, protected from the rigors of everyday use.

An easy experiment you can do to test the effects of abrasion on a coin is to place some dirt between your thumb and forefinger and rub it on a shiny penny for a few minutes. After washing away the dirt, you will likely notice scratches on its surface, as well as a noticeable decrease in luster. The next time you recover some change from the washer or dryer look for scratches on the coins—they might be clean, but the cleanliness comes at a cost.

If you spend any time at the beach, you will often see folks searching for hidden treasure with metal detectors. Sadly, any coins they may have been lucky enough to dig up will have been severely compromised by the sand and surf. Given enough time, waves can polish coins until they are smooth as sea glass.

When I was a kid, a perpetual item on my Christmas list was a rock tumbler, a gift I never received. When I became old enough to purchase one myself I decided against it when I found out how long it took to actually polish a rock. The basic idea is to tumble your rocks with varying grades of grit, starting with coarse grit and eventually moving up to the finer stuff. There are several steps, with each step typically taking seven days. I was curious as to whether a rock tumbler could be used to clean coins, or perhaps speed up their aging, so I purchased one and got to work.

I grabbed a handful of money from my coin jar, about sixty coins in all, added the coarse grit and water, and then set it up to tumble. The motor turned but the drum wouldn't spin. Evidently coins are a lot heavier than rocks. I took about half the coins out. Still no luck—it groaned but it wouldn't spin. I finally reduced the number of coins to around ten, at which point it spun unencumbered.

Two days later I opened up the drum, rinsed off the black gunk and was amazed by what I saw. All of the coins looked faded and bleached, as if they had weathered a lifetime of storms. They reminded me of the stone-washed jeans of the 1990s, which are literally stone washed, being tumbled about in heavy-duty washing machines with rocks.

The surface of each coin was so abraded that the luster was completely gone. They felt rough to the touch, as they were badly scratched. The copper layer on the pennies had worn considerably but not all the way through—its surface was pitted with tiny black dots, exposing the zinc core. The raised areas of the coins were much more pitted than the smoother areas. On the penny, Lincoln's head had received the most pitting. On the reverse side, the raised Lincoln memorial experienced significantly more pitting than the other areas which were smoother. Oddly, the surface of the coin felt simultaneously rough and smooth— smooth because the relief had worn down so much, but rough because of its pitted surface. The edge of the coin was gray, the copper having been completely rubbed off.

The silvery-colored clad coinage did not fare much better— their once shiny surfaces had been reduced to an ugly gray pallor. There was not a trace of corrosion on any of the coins—it had been thoroughly beaten off. Unless your goal is to permanently mar and disfigure your coins, no sane person would ever clean their coins with a rock tumbler. You will no doubt get your coins clean, but only because you are removing the outer layer of metal along with the dirt and corrosion.

I was curious as to what the coins would look like after a full week of abuse so I put the coins back in and let them spin for the rest of the week.

When five more days had passed I removed the coins. Their appearance was worse than I expected—the erosion of their surfaces had continued unabated. On one of the plated zinc pennies, the outline of Lincoln was almost completely black— the zinc core being ruthlessly exposed. There was a lot of pitting on all the pennies, again more prominently on the raised areas. The surfaces of the dimes and the one quarter were still a dull gray, but surprisingly had very little pitting, testifying to the toughness of the outer cupronickel layer on these clad coins.

There are numerous grading systems that have been devised by numismatists to rank coins, based largely on the amount of wear, and for nicer coins, their luster. The European grading system for coins contains eight categories (see Table 4.1), as follows:

Table 4.1   The 8 categories of European grading system for coins.

Designation	Original design remaining
Good	10%
Very good	25%
Fine	50%
Very fine	75%
Extremely fine	90%
About uncirculated	95% + some luster
Uncirculated	100% + some luster
Brilliant uncirculated or gem	100% + full luster

If you are a serious coin collector your goal is to obtain coins with the least amount of wear, as they will have the most value. When I was a kid, I would get excited when I came across an old coin. I would always be disappointed when I looked up its value, though, which would invariably be less than I had hoped, usually because the coin was severely worn. Most of the time we don't worry too much about the amount of wear or corrosion on a coin. Other than perhaps getting stuck in a vending machine, a worn or discolored coin is little more than a nuisance.

However, for much of history worn coins posed a much more practical problem. Except for fiduciary coins, the intrinsic value of a coin was typically based on the weight of the metal it contained. A worn coin would be worth considerably less.

The Coinage Act of 1870 set minimum allowable weight limits for each coin in circulation in Britain, not only to deter the common practice of shaving off precious metal from coins, but also to make it known when a coin was no longer legal tender, whether through legitimate wear and tear, or otherwise. The Act outlined the standard weight and the least acceptable weight of each type of coin. The allowances were quite paltry. The standard weight of a 5-pound gold coin was to be 39.940 g, while the least acceptable weight was 39.690 g. A half sovereign gold coin had to be between 3.961–3.994 g.

Passing underweight coins as legitimate currency was a serious offense, prompting the following directive in Section 7 of the Coinage Act:

> Where any gold coin of the realm is below the current weight, every person shall, by himself or others, cut, break, or deface any such coin tendered to him in payment, and the person tendering the same shall bear the loss.[1]

For a coin to retain the greatest percentage of its mass over time, it must not be too thin, as a thin coin has a small mass to surface area ratio. Thicker cylinders are preferable to thinner ones. The very thin coins of the late Byzantine era would be especially subject to wear. The largest mass to surface area ratio is found in a sphere. Until recently, the coin whose shape most closely approximated that of a sphere was the Siamese bullet money. These strange-looking silver objects were made from

Figure 4.1   Siamese bullet coins.
                © Morton & Eden.

1630–1780; they look as though someone has taken a hammer to both ends of a rectangular block of metal, as though they were trying to make a ball out of an ingot (Figure 4.1).

The world's first truly spherical coin was made by the Mint of Poland in 2015 for Niue Island, a territory of New Zealand. It was made from silver and featured the seven new wonders of the world. The coin is quite fetching—its intricate design is made possible by 3-D laser engraving. Many coin connoisseurs were not fans of the orb, though, claiming it looked more like a marble than a coin. Others likened its appearance to that of the *Star Wars* Death Star.

4.2   DIRTY COINS

As coins are handled they not only wear down but also get dirty. Common sense tells you that if a coin is dirty it needs a good washing. However, most coin experts will tell you never to clean a coin, as the cleaning process invariably does more harm than good. A trained eye can instantly spot an improperly cleaned coin. It will either be scratched or worn, severely diminishing its value.

If you must clean your coins, it is important to resort to non-destructive methods. To begin, use distilled water. Tap water contains chlorine, which can cause corrosion. Begin by just soaking your coins—don't rub or polish them. Let them air dry—don't use a cloth, not even a soft one. If you insist on scrubbing the dirt from a coin use a soft-bristled toothbrush, but don't even think about using toothpaste—the abrasives will scratch your coin's surface. (Tooth enamel is the hardest tissue in your body, being primarily made of hydroxyapatite, which comes in at 5 on Moh's scale, harder than most coins.)

If water doesn't clean your coin then a solvent such as acetone might do the trick. Acetone will dissolve a wide assortment of substances and is the active ingredient in nail polish remover.

Acetone is a good solvent for use on coins, as it has little effect on metal. Since acetone is somewhat polar, and as like usually dissolves like it readily dissolves in water. The carbonyl group ($C=O$) on one side of an acetone molecule is decidedly polar, while the two methyl groups ($-CH_3$) on the other side are non-polar. The polar side of acetone can dissolve other polar substances, while the nonpolar side is attracted to nonpolar species, giving acetone a great deal of versatility.

If acetone cannot remove the gunk on coins, the next thing to try is a mild soap solution using distilled water. The use of a detergent-based cleaner is not recommended, as detergents tend to be harsher, containing additional additives that may damage the coin. A good rule of thumb is to never use a soap on your coins that you wouldn't use on your hands. A molecule of soap has a nonpolar tail and a polar head; in some respects, it is similar to acetone in how it cleans. Soap is especially good for removing oily fingerprints from the face of a coin. When handling coins, it is always a good idea to grasp them by their edges as opposed to their faces, so as to minimize the amount of surface area which comes into contact with your skin.

Another substance hailed as a miracle worker for cleaning coins is olive oil. Olive oil is especially effective for coins that are heavily encrusted with multiple layers of hard-to-remove grime. Cleaning coins with olive oil is easy—just dump your coins in a container of it for a week, or a month, and let it work its magic. The nonpolar nature of the oil is quite effective at dissolving all variety of other nonpolar substances.

Some numismatists, however, are not enamored with olive oil, claiming that its fatty acids, primarily oleic acid, can damage coins. Extra virgin olive oil has the lowest concentration of free fatty acids, so would be the safest to use on coins; it is also the darkest, however, so can cause discoloration if coins are soaked in it for too long. The cheaper brands of olive oil are more acidic, containing a greater percentage of free fatty acids, potentially doing more damage to your coins.

Can a dirty coin look clean? To answer this question, pick up a handful of change and give it a good whiff. Coins have a distinctive smell, as do our hands after we have been jiggling around the change in our pockets. In 2006 a group of engineers from the Virginia Polytechnic Institute and State University, USA investigated coin odors. They discovered that it is only when the metal of the coin itself comes into contact with our perspiration that it develops its distinct metallic odor.

By analyzing the gases from the skin of those who touched iron, they were able to detect a specific type of volatile organic compound—1-octen-3-one—which has a fairly strong scent. They discovered that the lipid peroxides contained in human sweat react with Fe^{2+} ions to produce this smelly substance, which has an aroma somewhat reminiscent of mushrooms.[2] A similar mechanism has been proposed for other metals. Evidently coins themselves do not have a smell, but we most certainly do.

4.3 LUSTER

Perhaps the easiest way to distinguish a new coin from an old one is by its luster. Even the lowly copper cent can be regarded as a thing of beauty in mint condition, as copper is an exceptionally lustrous metal. Mineralogists have defined numerous types of luster to aid in mineral identification. Mica and pearls have a pearly luster, while amber is resinous; glass has a vitreous luster. Metallic luster refers to the luster that results from the reflection of light from its surface. Diamonds are also lustrous, not because they reflect light, but rather because they refract it—as light passes through a diamond it takes a circuitous path before it reemerges at a point far removed from where the incident light entered.

Metallic reflection is caused by electrons, specifically delocalized electrons. When a photon of light strikes the surface of a metal it is absorbed by one of these delocalized electrons, causing the electron to jump to a higher energy level. However, it only stays excited momentarily before falling back to the ground state. When it falls back, the previously absorbed light is reemitted—as a result, whenever light strikes the surface of a metal it tends to be reflected.

Transition metals are often the most lustrous, as their sea of free electrons is greatly enlarged owing to all of their delocalized d-electrons. Most of the transition metals look very much alike, being uniformly shiny in appearance.

In order for a metal to retain its reflective surface, however, it must have some degree of corrosion resistance. The shiniest metals are not necessarily those with the most original luster, but rather those best equipped to maintain it. Many elements that are very shiny lose their luster quickly when exposed to oxygen. If you cut open a piece of sodium or potassium with a butter knife you can see the shiny metal underneath, which quickly turns black owing to oxidation by the air. The more active the metal, the quicker this oxidation layer forms.

Chromium is generally considered to be a very shiny element—it is chromium that makes stainless steel stainless. Even small amounts of chromium added to other more active metals will make them corrosion resistant, and thus very shiny. High quality dental mirrors are usually plated with rhodium, as are those on the Space Station. Rhodium has excellent corrosion resistance, making it especially suitable for use as a reflector.

Two of our traditional coinage metals—gold and copper—technically are not as lustrous as silver as they do not reflect all of the light that strikes them; nevertheless, the nobility that their corrosion resistance confers has elevated them into the upper echelon of lustrous metals.

When it comes to coins, however, luster means a lot more than shininess. An uncirculated mint coin will possess a special kind of luster known as mint bloom, or mint luster. Mint luster can be observed when light interacts with the pattern of radiate flow lines that develop on a coin's surface after it is struck. When a coin is struck with a die to create an image, the metal undergoes a significant amount of strain, resulting in parallel striations radiating outward from the center of the coin. These lines reveal the flow of the metal as it was struck.

It is not uncommon for stresses to produce radial fractures—when a projectile pierces a pane of glass, for example, it will produce a series of radial fractures extending outward from the hole in the center from which the impacting force originated. Throw a pebble in a pond and you can observe the concentric ripples.

In the same way, when a coin blank is struck with a hammer and die it flattens out, because metal is so malleable the metal actually flows. However, since a coin is restrained by a collar during striking the metal cannot flow outward—instead it flows upward. As the metal still expands owing to the compressive force of the die it has no other place to go; each of the areas between the flow lines represents a raised area as the lines themselves are recessed. Over time, these striations are imprinted into the die itself, making the lines more prominent the more the die is used.

Some very conspicuous flow lines have been observed on a number of 2000-year-old silver and bronze Roman coins (Figure 4.2). The same dies were used repeatedly, therefore the flow lines were deeply etched into the coins—they are still visible after all these years. Today dies are polished frequently to prevent the flow lines from being too pronounced.

These striations are usually the most visible on the obverse side of a coin, which tends to be flatter than the reverse side. Some of the most striking radiate flow lines occur on the US. Morgan silver dollar of the late 1800s, a coin popular with collectors. Owing to its large size and relatively flat design, the mint bloom was easy to observe on these uncirculated coins. The Morgan dollar would eventually be replaced by the Peace dollar, which was minted to observe the end of World War I—these silver dollars also demonstrate exceptional radiate flow lines.

In 1964 almost 3 million Morgan silver dollars were discovered in a vault in the Treasury building in Washington, D.C. All were in perfect brilliant uncirculated condition, having been sealed in bags of a 1000 each, untouched for over 60 years. As they had never been circulated, the mint bloom was still present.

Figure 4.2   Ancient Roman Republic silver coin denarius 47-46 BC.
            © Shutterstock.

A trained eye can detect a counterfeit coin by noting whether or not the radiate flow lines are present. Coins that are not struck will not have these lines. Sometimes a genuine coin will be altered to increase its value. One coin that can easily be altered is the 1943 US steel penny. It is easy to electroplate an additional layer of zinc onto the steel core; doing so tends to fill in the radiate flow lines, though, revealing that the coin has been tampered with.

Mint luster is due to the interaction of light with these radiate flow lines. An interesting optical phenomenon, known as the cartwheel effect, can be seen on coins with mint luster. If held at an angle under a bright light and rotated in your hand, the radiating striations reflect light in such a way as to produce an alternating bright-dark pattern. It looks like two thin slices of pie opposite one another that are brightly reflecting light as they turn, as if they were doing cartwheels on a rotating platform. The cartwheel effect is so vivid it is even visible through transparent packaging over the coin. However, it can only be seen on uncirculated coins in mint condition—any attempts at cleaning will obliterate these lines, and with it the cartwheel effect. No matter how hard you try, once you remove a coin's mint luster you can't get it back. In attempting to distinguish between a cleaned and uncleaned coin, the cartwheel effect is one of the first things a trained eye will look for.

4.4 WHAT IS CORROSION?

If you were to take a new coin and put it in a drawer, would it remain in mint condition? Ask any coin collector, and you will be answered with a resounding "No!". Coins are under constant attack from gases in the atmosphere, and not just pollutants either—oxygen in particular is no friend to coins, and is capable of doing substantial damage. Atmospheric gases are among the chief causes of coin corrosion.

If your goal is to completely prevent corrosion, your best bet would be to store your valuables in a container evacuated of air and filled with a noble gas. A number of valuable documents and paintings are stored this way. However, not all noble gases are equal. The Declaration of Independence, one of America's most cherished documents, was stored in an atmosphere of humidified helium up until 2003, when the switch was made to argon. Humidity is necessary to keep old parchment from cracking, and as argon is more soluble in water than helium, it is easier to humidify. Helium atoms are also so small that they are hard to confine—consider how quickly a balloon filled with helium deflates. The heavier noble gases, such as krypton and xenon, are potentially more reactive; even though they are still quite inert, their valence electrons are far enough away from the nucleus (and thus not as tightly held as those in lighter, smaller noble gases) that they could potentially be coaxed into participating in some type of reaction. Therefore, argon is the best choice; it is also the easiest to obtain, making up 0.9% of the atmosphere.

Any metal exposed to the atmosphere is subject to corrosion. Corrosion occurs anytime a pure metal forms a compound—it is the reversion of a pure metal back to its natural, that is, the least energetic state. A system with the least amount of energy is the most stable. Chemists have devised their own special type of energy, known as Gibbs free energy, to refer to the amount of energy available to a system to get something done, that is, to do work. The less Gibbs free energy a system has, the more stable it is. In general, any system is more stable if it has less energy, regardless of the type.

Boulders roll downhill because of gravity, yes, but the driving force is a tendency toward a less energetic state—they have less gravitational potential energy at the bottom of the hill than at the top. Gasoline is quite unstable, containing a great deal of chemical potential energy. When combusted it turns into water and carbon dioxide, both very respectable, less energetic compounds. It doesn't take a lot of prodding to get gasoline to burn, but it would take quite a bit of alchemy to get water and carbon dioxide to turn back into gasoline.

Most of the pure metal we have today is mined as an ore. To retrieve the pure metal, this ore must be reduced, as discussed earlier. This reduction process is not spontaneous, however, as it requires the input of energy to occur. However, corrosion is a spontaneous process—it occurs without our meddling. Iron rusts all by itself, often despite our best efforts to prevent it.

The reverse of reduction is oxidation, therefore nearly every corrosion reaction involves oxidation, with a "goal" of returning the metal to its natural state. Every coin in our pocket is hell-bent on returning to its original form, which in most cases is a mis-shapen ugly lump of rock. Think of any refined metal as you would a ball rolling down an ever-so-gentle slope: unless you intervene, the ball will eventually find its way to the bottom of the slope. In the same way, every pure metal (with the possible exception of gold) is in the process of corroding, and will continue to do so until the corrosion process is complete. Keeping coins in good shape requires some effort—they can't be left to their own devices.

As almost every coin minted is made from some type of alloy or a combination of different metals, determining the exact chemical reactions that contribute to corrosion within any particular coin can be daunting. As each individual gas in the atmosphere can participate in their own distinct reaction with any specific metal within a coin, there will often be numerous reactions occurring simultaneously. Add in pollutants, that may or may not be present, and the picture becomes even more muddled.

A corroded coin is not the same thing as a dirty coin. Corrosion is not a layer of grime on the surface of a coin that can be washed off, but rather a compound made up partly of the coin's metal—if you remove this layer you are removing an outer layer of the coin. It is not possible to remove the corrosion product without removing some of the coin's base metal. If you wanted to remove some dirt from your body you would take a bath—removing a mole, however, is a bit more complicated.

4.5 PROTECTING COINS FROM CORROSION

One way to prevent corrosion is to store your coins in the proper container. However, just sticking your coins in a paper envelope is a no-no. If the paper contains acids, it can accelerate the corrosion rate of the coins. Plastics are often bad too; chlorine-based compounds like PVC are terrible for coins—you might as well just drop them in bleach. Even chlorine-free plastic is not good enough for really valuable coins; plastic is porous, and coin-corroding gases will find their way in, attacking precious coins with malicious glee.

There are a number of methods coin collectors use to protect their horde. One method used in the past, but not so much today, involves the use of volatile corrosion inhibitors (VCIs). The principle behind a volatile corrosion inhibitor (VCI) is very simple—a volatile chemical is placed in a sealed container along with the metal that is to be protected. The VCI can be incorporated within the material of the container itself, or added separately. As the VCI evaporates, it condenses a thin film of oil which adheres to the surface of the protected metal. Theoretically, this thin film will prevent atmospheric gases from reaching the metal, thus protecting it from corrosion. Any number of nasty organic compounds can be used for VCIs, usually an aromatic amine or some type of acrylate or nitrite compound.

A 1975 US patent application filed by Allan Sanford of Massachusetts, USA, lays claim to a device that will protect coins using VCI technology. In the patent supplication, Sanford makes the following assertion:

> The general objective is to provide containers for copper and silver coins, copper and silver composition coins, medals and like articles that will hold at least one such article and that contains a corrosion inhibiting agent, inert with respect to copper and silver and capable of affording protection for the stored article or articles over long periods of time.

He also provides the specific chemical composition:

> Suitable volatile corrosion inhibiting agents inert with respect to copper and silver are known. They include dicyclohexylammonium nitrite and also a solution containing 1.7% diisopropylammonium nitrite and 0.4% ethanolamine salicylate or 0.9% sodium mercaptobenzthiazole. The copper in structures consisting of steel and copper components and also silver may be protected, as disclosed by Metallic Corrosion Inhibitors in the presence of diisopropylammonium nitrite by adding 0.2% benzoin a-benzoin oxime, b-benzoin oxime, various isomeric napthylamine-sulphonic acids, 0-, m-, and p-phenylene-diammine, thiodiethylene glycol, the sodium salt of mercaptothiazdine or b- (2-mercaptobenzthiazy.l-) propionic acid.[3]

Despite the optimism expressed in this application, VCIs have largely fallen out of favor amongst coin collectors, as no one wants their coins coated in an oily film. The chemicals involved are not friendly to the environment or to human health either; significant tarnishing has also occurred as a result of storing coins in VCI containers. VCIs are still widely used in industry, though, being used mostly to protect sensitive metallic components during storage and shipping.

One intriguing piece of technology that has been developed to protect coins from corrosion involves embedding flakes of copper into coin boxes, albums and slipcovers. One popular brand is called *Intercept Shield*. Copper is used in flake form so as to maximize surface area. Instead of purchasing these products, some collectors will just place a shiny copper penny next to their valuable coins, roughing it up with some steel wool to increase its surface area. When it becomes tarnished it can be replaced.

The *Intercept Shield* technology relies on a neat chemical trick that will be familiar to any boat owner—the copper is acting as a sacrificial anode. If you examine any metal boat, chances are you will see strips of metal, often zinc, attached to its hull—these strips are acting as sacrificial anodes (Figure 4.3). Sacrificial anodes are especially important if a ship is to be used in salt water, as the salt can accelerate the rate of corrosion. If you examine the hot water heater in your home, you will likely see an indentation along the top in which a sacrificial anode rod (usually of magnesium) has been inserted. Sacrificial anodes

Figure 4.3 Group of sacrificial anodes on the waterjet of a fast boat in a shipyard for repairs and maintenance.
© Photomarine/Shutterstock.com.

are used on pipelines, water tanks, bridges or other structures in which metals are subject to corrosion; they do need to be periodically replaced, however, as they corrode at an accelerated rate.

The basic principle behind using a sacrificial anode is to protect a more valuable metal by placing a less valuable metal in harm's way, one that will be sacrificed for the greater good. In every case the sacrificial anode is a more electrochemically active metal—one more easily oxidized. As oxidation involves the losing of electrons, a more active metal is one that holds on to its electrons more loosely and is thus more likely to give them up. The activity series of metals is a listing of elements in order of ease of oxidation, and is ordered as follows:

Most Active (Most Easily Oxidized)
K
Na
Ca
Mg
Al
C
Zn
Fe
Sn
Pb
H
Cu
Ag
Au
Least Active (Least Easily Oxidized)

Those elements at the top of the chart give up electrons more readily than those toward the bottom. The activity series is quite useful in predicting whether or not a reaction will occur. Any element higher on the chart will replace an element lower on the chart. Earlier, we mentioned that copper does not react with hydrochloric acid but zinc does, as described by the following reaction:

$$2HCl_{(aq)} + Zn_{(s)} \rightarrow ZnCl_{2(aq)} + H_{2(g)}$$

The activity series explains why this reaction occurs. Copper is below hydrogen on the chart, but zinc is above it. Zinc is a more active element than hydrogen, therefore it gives up its electrons more readily. It takes less energy for chlorine to bond with zinc than it does hydrogen, therefore the chlorine will let go of hydrogen and instead bond with zinc.

To pose an analogy, it would be akin to a guy dumping his girlfriend for another girl who is decidedly more low maintenance. Copper, being near the bottom of the chart, is very high maintenance, whereas zinc is low maintenance. The chlorine atom, like the shallow guy, dumps its high maintenance partner (hydrogen) for one who is more low maintenance (zinc). At the very bottom of the chart is gold, who is so high maintenance that it seldom, if ever, can even find a partner.

By examining the activity series, you can see that using copper as a sacrificial anode to protect your coins is only going to work if they are made from silver or gold. Gold is so unreactive that it undergoes little corrosion, therefore the main purpose of using copper as a sacrificial anode is to protect silver coins. If you have some valuable copper coins, your sacrificial anode would need to be made of something higher than copper on the activity series, perhaps zinc. If your coin collection consists of random coins plucked from your pocket change then you probably don't need to concern yourself with sacrificial anodes at all.

If copper is the sacrificial anode when used with silver, then silver must be the cathode. We normally associate anodes and cathodes with batteries and circuits, in which electrons are flowing as electricity is generated. In essence, anytime an oxidation reaction is occurring, where electrons are being lost by one atom, a reduction reaction is also occurring, as those lost electrons must be gained by another atom. Therefore, when you throw a penny into your valuable collection of silver coins you are creating an electrochemical cell—more specifically a voltaic (or galvanic) cell. The heartbeat of a voltaic cell is a redox reaction. They occur spontaneously, needing no outside interference to occur.

4.6 DOES GOLD CORRODE?

Although gold is the most noble of the metals, gold coins can still experience tarnishing. As gold is often alloyed with copper, it is not unusual to see older gold coins experience some corrosion owing to the added copper. If you compare a new gold coin with one that is a hundred years old, the older coin will look darker. Occasionally a gold or silver coin will be found that has a green crust, which usually indicates copper.

Gold will not react with oxygen, but will react with chlorine, however. Gold is a metal, after all, and like all metals it will give up its electrons given enough prodding. Gold may be a noble metal, but it's not as noble as a noble gas. Even though oxygen is not electronegative enough to strip away gold's electrons, some halogens are. As the most common oxidation state of gold is $+1$, when it reacts with chlorine it forms AuCl.

In 2013, a California, USA couple was walking their dog when they happened to see a rusty can sticking out of the ground. The first thing they noticed was its heft. When they looked inside they saw it was filled mostly with 20-dollar gold coins—double eagles—dating from the 1800s. They would go on to discover seven more cans, eventually finding over 1400 coins. The coins themselves had a face value of \$28 000, but were valued at more like \$10 million. Despite having been underground for over a hundred years, most of the coins were in excellent shape—some were even in mint condition—although a few did show spots and discoloration.[4]

Gold coins recovered from shipwrecks are still in surprisingly good condition, despite prolonged exposure to seawater. Over a hundred gold coins recovered in 1988 from the wreck of the SS Central America, the "Ship of Gold", were still in excellent shape despite having been underwater for over 130 years. The ship was carrying 10 tons of gold when it sank off the east coast of the US upon encountering a hurricane, leading to the tragic loss of 425 lives. The large weight of gold most likely contributed to its sinking. Many of the recovered coins did show noticeable discoloration, most likely from the chlorine in sea water, but not nearly as bad as you might expect.

4.7 SILVER CORROSION

Silver that falls to the bottom of the sea does not fare nearly as well as gold—silver coins recovered from shipwrecks are often unrecognizable. Archaeologists from the Maritime

Western Australian Museum, who have successfully restored thousands of encrusted coins, provide this firsthand account of how silver coins often look when dragged from the depths of the sea:

A typical silver coin from these wrecks consists of a central core of uncorroded metal surrounded by a layer of pre-dominantly silver compounds, which we refer to as the corrosion layer. The major components of this layer are silver chloride (cerargyrite), a mixed silver chloride-bromide, and silver sulphide (argentite). The proportions of these compounds vary widely and are probably related to site conditions. On top of the corrosion layer is the con-cretion layer which typically consists of shell fragments, sand, copper compounds (from the copper in the original metal), iron compounds from other objects on the wreck site, and sometimes large amounts of silver sulphide. In some coins the corrosion has gone to completion and no solid core remains. In other cases, local conditions have resulted in the formation of metallic silver in the corrosion layer.[5] (Excerpt reproduced from ref. 5 with permission from Taylor and Francis, Copyright 1979.)

When removing these encrusted silver coins from the site of a shipwreck, extreme care must be exercised, as they can literally crumble in your hands when removed from the water. Silver is especially susceptible to attack from sulfur. Sulfur is silver's Achilles's heel. In seawater, sulfur is primarily in the form of sulfate ions, and even though the concentration is only around 2.6 mg L^{-1}, it is enough to cause significant tarnishing.

Silver is just as susceptible to sulfur in the air. One of the most common sulfur-containing gases is hydrogen sulfide (H_2S), which is toxic at even low levels. Exposure to a concentration as low as 800 ppm for 5 minutes is considered lethal. Hydrogen sulfide gas has a rotten egg odor, being detectable by the human nose in concentrations as low as 0.5 ppb (parts per billion). When it reacts with silver it produces a black film of silver sulfide, as described by the following equation:

$$2Ag_{(s)} + H_2S_{(g)} \rightarrow Ag_2S_{(s)} + H_{2(g)}$$

I still remember a story one of my grad school professors shared in class about the effects of sulfur on silver. I was not able to find any accounts of this incident in the literature (and the professor has since passed), so the account may very well be apocryphal, but it is simply too good not to share. An art museum, the story goes, had an impressive silver sculpture that was moved from storage to a prominent location in the museum's lobby, where it could be viewed by visitors as they waited in line. Museum workers noticed that the sculpture was tarnishing at an alarming rate after having been moved to this new location. Puzzled by this quickening pace of corrosion, they consulted a chemist. After running some tests, he concluded that the accelerated rate of tarnishing was a result of an increased concentration of hydrogen sulfide gas in the immediate vicinity of the sculpture, an increase he attributed to the emission of sulfur-containing flatulence from the visitors as they waited in line.

Any sulfur-containing compound can create tarnish. If using silver around foods that are rich in sulfur, such as eggs, broccoli or onions, the rate at which tarnish forms will increase. Curious as to the effect of sulfur-containing foods on the corrosion rate of silver, I placed a 1944 American quarter, which is 90% silver, in a baggie of diced onions. Within a day there was a noticeable blackening on the coin. After a week the coating was so thick the date was obscured. Upon removing the coin, I removed a piece of onion that was stuck to its surface; underneath the onion I noticed the coin was hardly tarnished at all. I found it intriguing that coming into contact with a piece of onion would protect the coin from corrosion rather than accelerate it.

I wiped the coin with a paper towel and was surprised at how much of the tarnish came off. However, after rinsing with water, much of the tarnish stayed put, covering the coin in irregular black splotches. The corrosion layer was very smooth to the touch, a little smoother even than the less tarnished parts of the coin. It even had a bit of a shine to it—one could almost describe it as lustrous.

Even though alloying silver with other metals will often decrease the incidence of tarnish, no silver alloy is completely corrosion resistant. Other metals, such as copper and brass, are also subject to the corrosive effects of sulfur-containing gases. If you examine any older silver coins, you will likely notice a black coating on the surface. If you examine the coins in your pocket, probably none of which contain silver, there is a good chance you will still observe some discoloration as a result of exposure to a sulfur-containing compound.

Water will greatly accelerate the tarnishing of silver (helping to explain why silver coins recovered from seawater fare so badly); it is thought that these corrosion reactions only occur on a thin layer of water on the silver's surface. High humidity will thus accelerate the tarnishing of silver.

Sulfur is not the only element that affects silver—chlorine is an even more effective oxidizer. Chlorine is ubiquitous—it can be found in cleaning products, in tap water and in the salt in our sweat, providing a good reason not to handle silver coins. However, as silver chloride is white, its presence is usually obscured by that of the darker silver sulfide.

Other gases can also have an effect on silver. Whereas silver does not readily react with atmospheric oxygen (O_2), it will react quite eagerly with atomic oxygen (O). If free oxygen is not available, a silver atom will do the next best thing and pluck one from a compound, explaining why silver reacts so readily with ozone (O_3). Ozone is responsible for the distinctive electrical smell in the air after a thunderstorm, and forms when elemental oxygen reacts with O_2. The elemental oxygen arises when lightning breaks apart O_2 into separate atoms. Silver has trouble removing an oxygen atom from O_2, however, as the bond strength of diatomic oxygen is just too great. The reaction between silver and ozone is as follows:

$$2Ag_{(s)} + O_{3(g)} \rightarrow Ag_2O_{(s)} + O_{2(g)}$$

As silver oxide is black its presence is often mistaken for that of hydrogen sulfide.

4.8 REMOVING SILVER TARNISH

Meticulous housekeepers often go to great lengths to remove silver tarnish. Commercial silver cleaners contain abrasive compounds such as chromium oxide, aluminum silicate or calcium carbonate that will remove this layer of corrosion. The problem with using abrasive compounds to remove tarnish is that silver is removed also, as tarnish is generally composed of silver sulfide. If you polish your silver enough, you will eventually end up with no silver at all!

There are several ways to remove unsightly silver sulfide from a coin without causing too much harm. The key is to find a substance that breaks down the silver sulfide coating—either through dissolving or chemically reacting—without damaging the underlying silver. Silver sulfide is known for its insolubility—its K_{sp} (solubility product constant) is an impressive 6.3×10^{-50}. K_{sp} values provide a way to quantify the solubility of a substance in water—the smaller the K_{sp} value the less soluble it is. By contrast, sodium chloride, which is quite soluble in water, has a K_{sp} value of 37.

Silver sulfide is insoluble in most acids and bases, too. It will dissolve in nitric acid though, but that's the last thing you want to use, as nitric acid reacts vigorously with silver, as well as copper. Silver sulfide will, however, react with alkali cyanide solutions, but cyanide compounds won't react with elemental silver, which is good. It is extremely important not to expose any cyanide compound to an acid, as doing so will liberate extremely toxic hydrogen cyanide gas.

A world-renowned numismatist died while cleaning silver coins with a cyanide solution, but not from the production of toxic fumes. In 1922, J. Sanford Saltus, president of both the New York Numismatist Club and the British Numismatic Society, was found dead in his London hotel room. He had prepared some cyanide solution in a glass, and was using it to clean some silver coins in his room. He had ordered some ginger ale, which he poured into a glass as well. Tragically, he drank from the wrong glass, dying shortly thereafter. Although the death was ruled an accident, others suspected suicide, as he was having some personal troubles—his fiancé had recently called off their engagement. Nevertheless, the official cause of Saltus's demise was listed by the coroner as "death by misadventure".[6]

Despite their toxicity, solutions made from sodium or potassium cyanide are highly effective at removing silver tarnish. A quick dip in a cyanide solution will remove silver tarnish almost immediately, as described by the following reaction:

$$Ag_2S_{(s)} + 4KCN_{(aq)} \rightarrow K_2S_{(aq)} + 2KAg(CN)_{2(aq)}$$

Both products, potassium sulfide and potassium silver cyanide, are water soluble, so rinsing the coin with water will remove these byproducts. Despite its effectiveness, cyanide solutions have somewhat fallen out of favor amongst coin collectors owing to their tendency to react with the silver, albeit slightly, producing a dull, cleaned appearance, stripping away some of the coin's luster.

Another method that produces good results is the acidified thiourea (NH_2CSNH_2) dip. Acidified thiourea is the active ingredient in Tarn-X Tarnish Remover. Thiourea is a chelating, or sequestering agent. Chelating agents have the ability to remove metals, and are often used to treat heavy metal poisoning. In cases in which a patient has been exposed to potentially harmful levels of heavy metals, such as mercury or lead, a chelating agent will be administered, which will bond to the toxic metal and form a complex, which can then be safely excreted from the body without doing further harm, as complex ions are usually water soluble.

Thiourea has the ability to remove, or sequester, the silver ion from silver sulfide by surrounding it, forming a silver complex in the process; hydrogen sulfide gas is also produced. The reaction is as follows:

$$Ag_2S_{(s)} + 2H^+_{(aq)} + 6NH_2CSNH_2 \rightarrow 2[Ag(NH_2CSNH_2)_3]^+_{(aq)} + H_2S_{(g)}$$

The problem with the thiourea dip is that even though it effectively removes silver sulfide from the surface of a coin, it is impossible to remove all of the thiourea from the coin, and a thin film will remain, causing eventual further corrosion as the sulfur reacts with the silver.

If you observe the above equations, you will notice that silver products are present in both. These methods have the unfortunate side effect of removing part of the valuable metal, providing yet another reason why any attempt to remove corrosion is so disdained by purists.

A much more effective way to remove silver tarnish, without losing any silver, is to perform a simple chemical reaction that reverses the process, which can be done quite easily using materials readily available from the grocery store. The first step is to take a piece of tarnished silver and wrap it in aluminum foil. It is then heated in a pan of water containing some baking soda (sodium bicarbonate). After a few minutes, the silver will be tarnish free. The reaction that occurs is:

$$3Ag_2S_{(s)} + Al_{(s)} \rightarrow 6Ag_{(s)} + Al_2S_{3(s)}$$

This equation represents a single displacement reaction, in which the aluminum takes the place of the silver, bonding with sulfur to yield aluminum sulfide, leaving behind elemental silver. As aluminum is higher on the activity series, it gives up electrons more readily than silver. As it is more energetically favorable for sulfur to bond with aluminum, the sulfur exchanges silver for aluminum, leaving behind silver in its pure form.

The reaction can also be described in electrochemical terms. As the silver and aluminum are touching one another, electrons spontaneously flow from the aluminum to the silver, creating a galvanic cell with a measurable voltage. As silver tarnish contains Ag^+ ions, the addition of electrons reduces it into pure silver. The baking soda adds some basic ions to the surrounding solution, which serves as the electrolyte.

4.9   TONING: CORROSION'S BEAUTIFUL SIDE

Corrosion is not necessarily viewed as a bad thing by coin collectors. Desirable corrosion even has a special name, being referred to as "toning." Toning adds color and character to older coins—many beautifully toned coins will exhibit all the colors of the rainbow. However, too much toning is not a good thing either, so while most collectors don't necessarily go looking for corrosion, they have learned to embrace it when it inevitably arrives. Although toning can appear on a variety of coins, it is most prevalent on those made of silver or copper. Many alloys used in coins today are so corrosion resistant that toning is not often seen; cupronickel, for example, is especially corrosion resistant.

The work of chemist Weimar W. White has contributed significantly to our current understanding of how corrosion affects coins. White meticulously measured the thickness of sulfide corrosion on older US quarters composed of 90% silver and 10% copper. He began by cleaning off the existing layers of corrosion and then immersing the clean coins in a sodium sulfide solution for various lengths of time. Both the silver and copper formed sulfide layers, as the following reactions show:

$$4Ag_{(s)} + O_{2(g)} + 2H_2O_{(l)} + 2Na_2S_{(aq)} \rightarrow 2Ag_2S_{(s)} + 4NaOH_{(aq)}$$

$$4Cu_{(s)} + O_{2(g)} + 2H_2O_{(l)} + 2Na_2S_{(aq)} \rightarrow 2Cu_2S_{(s)} + 4NaOH_{(aq)}$$

After the reactions, he used several different methods to determine the mass of these sulfide layers. From the mass, he was able to determine the thickness of each layer. White discovered a typical sulfide layer to be anywhere from 40–125 nm thick. (A piece of paper is about 100 000 nm thick.) White noticed that different thicknesses of toning yielded different colors. The thinnest layers were yellow, and as the sulfide layer got thicker, it turned red, then purple and finally blue. The thickest layers were black. A deposit of 4 $\mu g\,cm^{-2}$ produced the yellow toning, 7 $\mu g\,cm^{-2}$ produced red toning, and blue toning was produced by a thickness of 11 $\mu g\,cm^{-2}$.

By using Avogadro's number (the number of atoms in a mole—6.022×10^{23}) and the percentage composition of the silver and copper sulfide products, White was even able to calculate the number of sulfur atoms per square centimeter on toned coins of different colors (Figure 4.4). There were 75 quadrillion (7.5×10^{16}) sulfur atoms in a yellow-toned coin, 131 quadrillion in a red-toned coin, and 206 quadrillion in a blue-toned quarter.[7]

Figure 4.4 Toning on a "Morgan" US silver dollar. Image courtesy of PCGS (www.pcgs.com/cert/36906135).

The issue of toning has been the subject of considerable debate amongst coin enthusiasts. Some argue that toning disfigures the natural state of the coin and should be avoided. Most, though, feel that toning is a natural process that enhances a coin's value, lending an authentic vibe to the coin. What everyone can agree on is that if a coin is toned, it should not be removed, as doing so will reduce its value.

On the other end of the spectrum are those who intentionally tone coins, usually by exposing them to a sulfide solution. This practice of intentional toning has been criticized by purists who feel it's not "natural". Others argue that the chemical composition of an intentionally toned coin is exactly the same as that of one that has been toned naturally. What is frowned upon, however, is the intentional toning of a coin to cover up defects or excessive wear—this practice is considered unethical.

The vivid coloring on a toned coin is a result of thin film interference, the same phenomenon that creates rainbows on a soap bubble or a thin film of motor oil atop a mud puddle. Whenever a light wave encounters any medium, it can be reflected, refracted or transmitted, or some combination thereof. Black surfaces can be shiny or dull—shiny black reflects more light than dull black; and even though both colors lay claim to being "black" they are not absorbing the same wavelengths.

Thin film interference is only present if the thickness of the film is very close to the wavelength of visible light, somewhere between 100–1000 nm. When white light strikes a thin film, some wavelengths bounce off while other wavelengths penetrate it. Depending on the substance, certain wavelengths will pass through but others won't. It's not hard to tell which wavelengths of light are reflected—the reflected wavelengths are the ones that produce the colors we see. Those wavelengths that do pass through a thin film will eventually reach the surface of whatever the film is resting upon. At this point, the waves may continue on their journey, or they may be reflected or refracted yet again.

If the waves that are reflected from the bottom of the film are close enough to the waves that are reflected off the top of the film, and in phase with these waves (their crests and troughs lining up with one other), then these waves will constructively interfere and be strengthened, producing a color that dominates the other wavelengths. If waves that produce yellow light constructively interfere, then yellow light will be evident. If the film is thicker in one spot than in another, then several different colors of light may be evident, producing a rainbow pattern. If the entire film appears black, then it is likely too thick to allow for this type of interference—not enough light is reflected back to produce any discernible color. If viewing multiple colors, you will typically see bands where color is absent—these regions represent areas where the waves destructively interfere.

Seldom will a toned coin be a uniform color. The relief on the surface of any coin will invariably affect the thickness of the corrosion layer that forms on a particular part of its surface. Although some types of metals buck the trend, generally the rougher the surface the more likely it is to corrode. A rougher surface has a greater surface area, making it more likely to react with any incoming oxidizers, a phenomenon readily observed in a multicolored toned coin. Often the yellow toning, which is a result of the thinnest layer of corrosion, will appear on the smoother parts of the coin, while the blues and purples appear on the areas of greatest relief. There is tremendous variability, however, in what parts of the coin will exhibit a particular color of toning. There are other factors besides roughness that can accelerate corrosion. Microscopic defects on a coin's surface can provide a foothold for corrosion. If parts of a coin are dirty, these areas may either protect the coin or cause it to corrode faster, depending on what the "dirt" is.

4.10   THE "TONING" EFFECT OF OUR FINGERPRINTS

If you want to tone your coins really fast just touch them. Fingerprints are a ready source of corrosion, wreaking havoc on a coin's surface—visible corrosion can sometimes be seen on a coin just a day or two after being handled with bare hands. Fingerprint residue is a potent soup of toxic compounds, containing fatty acids, wax esters, proteins, cholesterol and a variety of inorganic salts.

If you're a smoker, nicotine shows up in your fingerprints, and if you use cosmetics regularly traces of it will appear as well. The composition of latent fingerprints is very similar to the composition of sweat. Sweat contains a variety of sulfur-containing compounds, including several types of sulfanyl alcohols, which contain both a hydroxyl group (–OH) and a thiol group (–SH). Sulfur is every coin's archenemy, as is chlorine, which is abundant in the sodium chloride found in the salt of sweat.

One group of Mexican researchers did a study to determine the effects of sweat on the corrosion rate of coins, immersing twenty copper-based coins of varying composition in a solution of artificial sweat for 24 hours. Significant corrosion was noted, with the two most common corrosion products being copper(I) oxide (Cu_2O) and copper(II) chloride hydroxide ($Cu_2(OH)_3Cl$), demonstrating that both oxygen and chlorine play a key role in the corrosion process.[8]

4.11   THE TONING EFFECT OF COIN CONTAINERS

The next time you get a paper roll of coins from the bank, look closely at the two end coins for signs of toning—they will often experience significant toning owing to contact with the paper. Older coins are especially susceptible to corrosion from the containers used to store them—especially if the container is made from paper or cardboard. During the paper-making process, sulfur compounds are used to turn wood into pulp, which is then made into paper. If you drive past a paper mill you will smell it long before you see it; the foul odors being emitted are sulfur-containing gases.

So, it shouldn't come as a great shock to learn that residual sulfur compounds are left behind in paper. The popularization of scrapbooking has made "acid-free paper" a household term. Acid-free paper is really sulfur-free paper—during processing the pulp is treated with a bicarbonate product to neutralize the acid. A lot of paper today is acid-free, but not all of it. A simple chemical test for sulfur can let you know if your paper contains sulfur, and by extension, acid. By burning a sample of paper and exposing the fumes to a piece of moistened lead acetate test paper, you can determine if sulfur is present. If sulfur is present, the paper will turn black due to the formation of an insoluble lead sulfide precipitate, as described by the following reaction:

$$Pb(CH_3COO)_{2(aq)} + H_2S_{(g)} \rightarrow PbS_{(s)} + 2CH_3COOH_{(aq)}$$

To verify that the paper truly turned black as a result of the presence of sulfur and not just from the deposition of carbon from the ash, you can add a little hydrogen peroxide, which will turn the paper white. Hydrogen peroxide reacts with the sulfide compound to yield lead sulfate, a white crystalline compound that forms according to the following reaction:

$$PbS_{(s)} + 4H_2O_{2(aq)} \rightarrow PbSO_{4(aq)} + 4H_2O_{(l)}$$

Storing coins in a wood container can also be problematic. Coin cabinets should never be made from oak, pine, cedar or any type of wood that tends to outgas volatile organic compounds, which can wreak havoc on coins. The very thing that makes a cedar chest great for repelling moths makes it terrible for coins. Wood that has been treated with stains and varnishes only compounds the problem, as these coatings release their own gases. If you really feel the need to house your coin collection in a wooden cabinet, mahogany or rosewood can be used, as they tend to be relatively inert, with a low rate of outgassing. Therefore, if you have your valuable coins tucked away in a dresser drawer, make sure the wood is coin-friendly.

As polymers began to gain popularity in the mid-twentieth century plastic began to replace paper and cardboard as the material of choice for coin collections. However, as we have seen, many plastics also contain chlorine, which is not much better than sulfur—copper coins are especially subject to attack by chlorine.

If you have any coins of value, it is best to visit your local coin dealer and buy some sulfur-free, chlorine-free coin holders. If your goal is to tone your coins, however, then your best bet is to put them in the cheapest brown paper bag you can find.

4.12   PASSIVATION

Among the many reasons not to clean coins, perhaps the most compelling is that by doing so you will expose your coin to further oxidation, and thus further conversion of the base metal into a worthless compound. A layer of corrosion protects the underlying metal. Whenever the underlying metal is protected by a layer of surface oxidation passivation has occurred.

With some metals, it is advantageous to speed up the formation of this surface layer of oxidation through an electrolytic process, so as to offer the greatest degree of protection in the shortest possible time. Metals subjected to this process are said to be anodized, as the metal to be oxidized acts as the anode. Anodized aluminum is perhaps the most common example of a metal prepared this way. The aluminum metal is immersed in an electrolyte through which an electric current is flowing—a thick layer of aluminum oxide is quickly deposited on the metal's surface, much thicker than would form naturally.

It may seem surprising that corrosion can have such a beneficial effect. For a rather crude analogy, consider the protection that a suntan might offer you from the sun. For sure, getting a tan is hard on the skin—it probably started with a sunburn, and ideally, a person who never received a suntan would be better off in the long run. However, once you have a tan you are afforded excellent protection from the sun. You would be foolish to stay indoors until the tan faded before venturing out into the sun again. It would be even more foolish to try to remove your tan—you couldn't do so without taking some skin with it. Once you're tanned, you don't have to worry about getting burned anymore—your skin is protected from further damage (more or less). In the same way, a layer of corrosion will protect a metal from further oxidation. Not only does removing a corrosion layer damage the coin and reduce the mass of precious metal, it will invariably make it even more vulnerable to the very thing you are trying to prevent.

The oxide layer that forms does not protect every metal, however; for some metals it affords little protection. Metallurgists rely on the Pilling–Bedworth (P–B) ratio to determine if an oxide layer will cause passivity in the underlying metal. This ratio was devised by N.B. Pilling and R.E. Bedworth while doing research for Westinghouse Manufacturing and Electric Company on the corrosion rates of metals exposed to high temperatures. In 1922 they laid out the basic premise for their now famous formula:

> If the layer of oxide formed takes up more space than the volume of metal which was oxidized to produce it, the oxide layer forms a tight fitting, solid sheet; if the reverse is the case, there is more space available within the original metallic dimensions than the oxide density requires, and a porous bricky structure is the result. In the former case, the enveloping oxide layer may impede further oxidation; in the latter, by virtue of its porosity, it does not.[9]

They devised a formula known as the critical density ratio (now referred to as the P–B ratio), which can be used to determine if an oxide layer will protect the metal. They stated their original formula as:

Wd/wD

W = the formula mass of the oxide
w = the formula mass of the metal
D = the density of the oxide
d = the density of the metal.

As density is defined as mass/volume, multiplying mass by density gives you the volume, therefore the ratio can be rewritten as: volume of oxide/volume metal (per metal atom).

If the P–B ratio is less than 1, then the oxide layer is too thin and the underlying metal will not be protected. When a ratio is greater than 2 the oxide expands too much upon its formation and it will tend to flake off. A ratio between 1 and 2 is ideal for passivation. Pilling and Bedworth noted that the following elements had critical density ratios less than 1:

Sodium (0.032)
Potassium (0.51)
Lithium (0.60)
Strontium (0.69)
Barium (0.78)
Calcium (0.78)

Each of the above elements are light metals, from groups 1 and 2 on the periodic table; as they are characterized by a low density, it is not surprising that the oxide layer would be too thin to offer sufficient protection. None of these metals are suitable for coin making—it doesn't help that most of the alkali metals explode when they come into contact with water.

Elements in which the P–B ratio are in the sweet spot, whose oxides readily provide passivation to their metal, are as follows:

Aluminum (1.28)
Copper (1.70)

Nickel (1.68)
Tin (1.33)
Zinc (1.59)

Each of the above elements has at some time been used for coins, as the oxide layer provides a great deal of protection. The following metals were noted to have P–B ratios greater than 2, leading to poor passivation:

Iron (2.06)
Manganese (2.07)
Cobalt (2.10)
Tungsten (3.30)
Chromium (3.92)

Chances are you will not find a coin made from any of the above metals, at least not on the outside. When rust (iron oxide) forms, it flakes off as it expands, weakening the metal and thus failing to provide any protection, instead exposing the under-lying metal to even more oxidation as it peels off. As mentioned earlier, iron has only been used for coins if it is overlaid with another metal, one whose oxide provides passivation.

Pilling and Bedworth go on to explain the ramifications of the data they collected:

> At any time the rate of oxidation is inversely proportional to the thickness of the layer of oxide than on the surface. The fact that this rate varies quantitatively with the amount of oxide overlying the metal surface points to an important conclusion: that the speed of the reaction is not determined by any property of the metal at all, but is determined by some combination of physical properties of its oxide. . ... The way in which the oxide layer limits the rate of oxidation can only mean that the transmission of oxygen through this layer to the reaction surface, which is the juncture of metal and oxide, is effected against a resistance which is pro-portional to its thickness.[9]

There are exceptions, of course, in which metals do not behave in the way the P–B ratio says they should. Most of the research by

Pilling and Bedworth dealt with oxides, therefore their ratios do not always apply to other types of corrosion products. Nevertheless, their ratio does provide a useful rule of thumb, and if you are looking for a metal that holds up under corrosion you can't just look at the activity series—you must also consider whether or not its corrosion products will help it or hurt it.

4.13 COPPER'S DUAL OXIDATION STATES

Most coins are alloys of different metals; in a perfect world the various types of metals would all be mixed evenly, but this is seldom the case, especially with older coins. In a silver or gold coin alloyed with copper, the copper will always corrode the fastest, as copper is a more active metal. If you see spots of corrosion on a gold coin, it is almost always caused by areas of the coin in which the copper concentration is abnormally high.

As we have seen, copper is more reactive than either silver or gold. Silver only has a single oxidation state: +1. Gold can have a +1 (aurous) or +3 (auric) oxidation state, yet these compounds are rarer than pure gold. Copper, on the other hand, has two very common oxidation states: +1 (cuprous) and +2 (cupric). Copper's two oxidation states make it more versatile, increasing its reactivity.

Unlike silver and gold, copper readily reacts with atmospheric oxygen. Many forms of additional copper corrosion only occur after the neutral copper atoms have been oxidized into copper ions, making the surface more amenable to further attack. The two oxidation states of copper can be easily demonstrated by heating a copper penny over a Bunsen burner flame. The reddish color that forms on the coin is caused by the formation of copper(I) oxide (occurring naturally as the mineral cuprite). If you continue to heat the copper penny in a flame, it will eventually turn gray. Copper(I) oxide further oxidizes into copper(II) oxide (tenorite), a much more stable compound. The reaction is as follows:

$$2Cu_2O_{(s)} + O_{2(g)} \rightarrow 4CuO_{(s)}$$

If you rub off this thin gray layer, you can usually see the red layer underneath.

Another way to observe the two oxidation states of copper is by using Benedict's solution, which is prepared by dissolving copper(II) sulfate, sodium carbonate and sodium citrate in water. Benedict's solution is used to test for the presence of reducing sugars such as glucose, lactose and maltose. When a reducing sugar is added to Benedict's solution and heated gently, the sugar will reduce the oxidation state of the copper from +2 to +1, forming copper(I) oxide, which, being insoluble, settles out, forming a handsome brick red precipitate. If a piece of copper looks red, it has probably been oxidized into the copper(I) form. Pure copper is copper-colored, not red.

As far as coin corrosion goes, though, copper(I) oxide is not the major player. The darkening of copper coins is primarily caused by the formation of copper(II) oxide, although there are likely to be other forms of oxidation as well on the surface of any copper coin. As copper(II) oxide is fairly stable it has been put to use in a variety of applications, and is often the material of choice if a deep black hue is desired, such as in ceramics or fingerprint dusting powder.

It is fairly easy to remove a layer of copper(II) oxide from a coin, providing you resign yourself to the fact that it will forever be a damaged "cleaned" coin. However, if your goal is to simply return some luster to your change, there are a plethora of options available. Steel wool will take the oxide layer right off, but will also leave your coin full of scratches.

Copper(II) oxide is soluble in just about any acid, dissociating into Cu^{2+} and O^{2-} ions. However, many of these same acids will also react with the copper. Theoretically, hydrochloric acid should be the best acid with which to clean a copper coin, as it doesn't react with copper; however, there are few coins made from pure copper. American copper pennies minted before 1982 still contain 5% zinc, which reacts readily with hydrochloric acid. The newer pennies, which are mostly zinc with a thin copper plating, do not fare well in hydrochloric acid either—the copper layer is so thin that hydrochloric acid can usually manage to find a way through. The newer British pennies fare a little better, as they are 6% copper, with the rest being steel, a metal almost as active as zinc.

4.14 REMOVING COPPER CORROSION

As countless science fair projects have shown, soaking tarnished pennies in vinegar and salt will produce astounding results, producing shiny pennies overnight, or quicker if the corrosion is light. Using vinegar by itself is not that effective though, but vinegar and salt works wonders. The addition of salt to remove corrosion at first seems counterintuitive, as it is common knowledge that salt speeds up the corrosion of metals.

Any type of chloride salt added to a weak acid will facilitate the cleaning of copper—potassium chloride works just as well as sodium chloride; the metal ions are merely spectators. Other types of weak acids work just as good as vinegar in cleaning pennies—the acetate ion also being also a spectator. The chloride ion, though, is key, speeding up the reaction exponentially; surprisingly, the exact mechanism by which this occurs has still not been established with certainty, although various mechanisms have been proposed.[10]

In addition to being soluble in acids, copper(II) oxide is also soluble in ammonia, which I have found to be very effective at cleaning copper coins. If placed in household ammonia, tarnished pennies will be bright and shiny in no time. A side benefit of using ammonia is that you get to observe a separate reaction occurring in the solution—it will turn an intense blue color owing to the reaction of ammonia with copper, forming the tetraamine copper(II) complex ion. The reaction is as follows:

$$4Cu_{(s)} + 16NH_{3(aq)} \rightarrow [Cu(NH_3)_4]_4{}^{2+}{}_{(aq)}$$

The deep blue color is quite striking. Complex ions are often characterized by intense colors; another example is heme, the functional group that makes up hemoglobin, which gives oxygenated blood its bright red color.

Not all bases are effective at cleaning pennies, however, especially those that contain chlorine, which is not at all friendly toward copper. If pennies are added to household bleach, which is typically a 5% solution of sodium hypochlorite (NaClO), in a short time they will be covered with a greenish-white crust of copper(II) chloride. The bleach will readily break apart the copper oxide coating on the penny, and since it doesn't know when to stop it will commence immediately to attacking the copper. Being such a strong oxidizer, its appetite for electrons is voracious. The reaction of bleach with copper is as follows:

$$Cu_{(s)} + ClO^-{}_{(aq)} + H_2O_{(l)} \rightarrow Cu^{2+}{}_{(aq)} + Cl^-{}_{(aq)} + 2OH^-{}_{(aq)}$$

4.15   COLORFUL COPPER CORROSION COMPOUNDS

A good place to view the oxidation of copper is at the Andy Warhol Museum in Pittsburg, PA, USA. On display is a piece of artwork titled appropriately enough *Oxidation*, consisting of metallic copper paint applied to canvas, which has been oxidized by the application of urine. The flow and drip patterns created by the streaming liquids are clearly visible; their reaction with copper creating a striking tapestry of various hues. The bluish-green hues are especially impressive; other parts of the oxidized surface are almost white, while others are black.

Warhol created these "paintings" by inviting friends over to urinate on large copper-coated canvases, letting nature take its course from there.

In an acidic environment and in the presence of oxygen copper will oxidize into Cu^+. Urine has an average pH of around 6.2. Each copper atom gives up an electron, which is accepted by oxygen, causing it to be reduced. The redox half-reactions are as follows (the hydrogen ions are supplied by the acidic urine):

$$\text{Oxidation: } Cu_{(s)} \rightarrow Cu^+_{(aq)} + e^-$$
$$\text{Reduction: } O_{2(g)} + 4H^+_{(aq)} + 4e^- \rightarrow 2H_2O_{(l)}$$

In the presence of a salt, the copper(I) ions quickly oxidize into copper(II) ions, forming copper(II) oxide, creating the black colors on the *Oxidation* paintings. The green colors are caused by the reaction of copper with chloride ions, with the chloride being supplied by the salty urine.

If a coin has turned blue or green, or something in between, it is almost certainly caused by the oxidation of copper, as copper forms a plethora of inorganic salts in the green-blue part of the spectrum. Studies of medieval artwork have shown that copper salts were the basis of many of the most vivid blue and green pigments used in paint. Beautiful green pigments were obtained from atacamite ($Cu_2Cl(OH)_3$), while posnjakite ($Cu_4SO_4(OH)_6 \cdot H_2O$) was used to create stunning blue pigments. Other pigments were made from copper acetate, malachite and azurite.

Copper(II) acetate is one of the easiest forms of copper corrosion to replicate. If some copper pennies are placed on a paper towel soaked in vinegar, copper(II) acetate will form overnight, forming a lovely blue-green patina. However, if copper pennies are simply immersed in vinegar, little will happen. In order for copper acetate to form, it must first react with oxygen to form CuO, after which it reacts with the acetic acid in vinegar to yield copper(II) acetate. The reaction is as follows:

$$CuO_{(s)} + CH_3COOH_{(aq)} \rightarrow Cu(CH_3COO)_{2(s)} + H_2O_{(l)}$$

There are so many greenish copper compounds that keeping track of them all can be overwhelming. Numismatists get around the confusion by referring to any green copper compound as verdigris, which translates from the French to "green of Greece". (The Greeks were among the first to make green pigments from verdigris.) Verdigris usually refers to compounds of basic copper(II) carbonate, or copper(II) acetate, but can be used for any type of green oxidation involving copper.

If you have a greenish corrosion layer on a copper coin, and you are curious as to its identity, one way to find out is to determine what type of solvent will remove it. If alcohol takes it off it could be copper(II) acetate, which is soluble in everything from rubbing alcohol to ethanol. If alcohol doesn't work, you can try an acid—basic copper(II) carbonate is soluble in most acids. If you think your coin is valuable, though, don't dip it in any liquids. Ancient copper coins that are coated with a green patina are prized by collectors and removing it will seriously compromise its value.

Copper corrosion forms the lovely green patinas that coat the surfaces of copper-containing statues, roofs, or outdoor furniture. A patina can refer to any layer of corrosion that forms on the surface of an object, metal or otherwise, while verdigris refers to the actual pigment (actually, the distinction is murky). Regardless, the term *patina* has come to be associated with the common blue-green corrosion so often seen on copper metals. This green color is typically the result of copper reacting with carbon dioxide in the air to form copper(II) carbonate ($CuCO_3$), but it can also arise from other sources. In polluted urban areas, sulfate-containing patinas are quite common; in locations close to the ocean chloride-containing patinas may predominate. The Peckham Library in London, NEMO Science Center in Amsterdam, and the Nordic Embassies in Berlin are each quite striking owing to their exteriors of copper-containing patinas.

I once purchased a set of Israeli coins from the bargain bin of an antique dealer. They came in a decorative cardboard holder covered with transparent plastic, and were advertised as "proof-like issues" from the Tel Aviv Mint, dating from 1966–67. I purchased them not because they were in mint condition, but because they were expressly *not* in mint condition—each coin was showing significant corrosion, even though they were still tightly ensconced within their packaging.

Five of the six coins had noticeable verdigris along their edges, which was gradually creeping inward. The three bronze-colored coins were showing the most corrosion; interestingly, the majority of the corrosion began on the same side of each coin—the left side. The 5 and 10-Agorot coins had most of the corrosion on the obverse side, but the reverse side of a larger 25-Agorot was the most corroded. As all of these coins contained copper, with smaller amounts of tin and aluminum, the greenish corrosion was not a surprise.

There were two silver-colored coins, a 1/2-lira and 1-lira, each with a cupronickel composition (75% copper and 25% nickel). This corrosion was a darker green in color, and not as pronounced as that of the bronze-colored coins.

There was one coin that stood apart from the others, a little 1-Agora aluminum coin, weighing in at a gram. It was the only one in the set that was not green, instead being covered by a layer of white corrosion completely obscuring the details of the coin. This amount of corrosion was surprising, as aluminum is usually protected from corrosion by its surface layer of aluminum oxide. I had a hunch that the corrosion product was aluminum chloride, which is also white. The chlorine would have come from the plastic used in the packaging, which was most likely made from PVC (polyvinyl chloride), or some other type of chlorine-based plastic.

The plastic of choice for coin containers today is polyethylene terephthalate, commonly known as PET (or as a film you might know it as Mylar), which is identified by recycling code #1. It is commonly used in 2-L soda bottles, which are perfectly safe for storing coins, providing you first rinse out the acidic soda residue.

4.16   THE STATUE OF LIBERTY, A CASE STUDY IN CORROSION

The Statue of Liberty has stood for over a hundred years in New York Harbor. When it was first given to the Americans by the French, it was copper-colored. Over time it became more reddish as the copper oxidized into copper(I) oxide. It eventually turned a dark brown, owing to further oxidation of the copper(I) oxide into copper(II) oxide. Over time Lady Liberty's 30-ton copper skin would turn bluish-green. This skin is only as thick as two US pennies, but if melted down would make over 400 million pennies. There are at least three reasons for this bluish-green coloration. First, the copper(II) oxide layer most likely reacted with carbon dioxide in the atmosphere to form a green layer of basic copper(I) carbonate, also known as malachite, according to the following equation:

$$2CuO_{(s)} + CO_{2(g)} + H_2O_{(l)} \rightarrow Cu_2CO_3(OH)_{2(s)}$$

In a more acidic environment, carbon dioxide can react with copper(II) oxide to form a blue layer, known as azurite, as shown by this equation:

$$3CuO_{(s)} + 2CO_{2(g)} + H_2O_{(l)} \rightarrow Cu_3(CO_3)_2(OH)_{2(s)}$$

Over time, azurite will usually turn into malachite.

Finally, sulfur dioxide (SO_2) pollutants in the air can also react with copper. SO_2 reacts with oxygen to form sulfur trioxide (SO_3), which reacts with copper(II) oxide to yield bluish-green brochantite ($Cu_4SO_4(OH)_6$), adding yet another component to the patina, according to this reaction:

$$4CuO_{(s)} + SO_{3(g)} + 3H_2O_{(l)} \rightarrow Cu_4SO_4(OH)_{6(s)}$$

In the 1980s, the Statue of Liberty underwent an extensive renovation. Surprisingly, it was discovered that it had only lost, on average, a tenth of a millimeter of copper from its surface. Even though it had been subject to the elements for a century, its extensive green and blue patina did not compromise the statue but rather protected it, providing a wonderful example of the passivating effects of corrosion.

The most disturbing aspect of the statue, which initially spurred the renovation effort, was hidden from view. The internal iron framework had become so badly corroded that it was becoming structurally unstable. Unlike the copper coating, which was firmly protected by a rich layer of verdigris, the inner skeleton was becoming seriously compromised. The culprit was bimetallic (or galvanic) corrosion, which occurs when two dissimilar metals come into contact. When it was built, asbestos insulators were inserted between the copper and iron, but over time much of this insulation had worn away, causing the iron and copper to come into direct contact. The salty mist blowing in from New York harbor created the perfect electrolyte, essentially converting the statue into a giant 46-meter tall battery.

As iron is more active than copper it gives up its electrons more easily. As electrons flow from the iron to the copper, the iron is oxidized, corroding at a faster rate. The oxidation half-reaction is as follows:

$$Fe_{(s)} \rightarrow Fe^{2+}{}_{(aq)} + 2e^-$$

The electrons released by the iron flow towards the copper and are picked up by dissolved oxygen in the water, as shown by the following reduction half-reaction:

$$H_2O_{(l)} + \tfrac{1}{2}O_{2(g)} + 2e^- \rightarrow 2OH^-{}_{(aq)}$$

The iron then reacts with these hydroxide ions to form iron(II) hydroxide, which is further oxidized to form the more familiar rust known as iron(III) hydroxide.

4.17 BIMETALLIC CORROSION

Bimetallic corrosion is quite common, occurring when two dissimilar metals come into contact. If an electrolyte gets between the metals, the corrosion rate speeds up considerably. If you are putting a nut on a bolt it is important they are made from the same metal—if not you will likely see bimetallic corrosion. If the bolt is placed against a different type of metal, you will eventually notice some discoloration where the two metals meet. If you have metal dental fillings, biting down on a piece of aluminum foil will enable you to experience bimetallic corrosion firsthand—the slight tingling sensation you feel results from electrons flowing from one metal to another, the saliva providing the perfect electrolyte.

Bimetallic corrosion can seemingly occur any time two disparate metals come into contact. The Mokume Gane (pronounced moe-koo-may-gah-nay) technique is used to make an exceptionally beautiful type of jewelry that relies on a traditional Japanese method of fusing together differing types of metal into a flowing pattern that resembles wood grain (Figure 4.5). The term *Mokume Gane* translates into "wood grain metal". Mixing copper and silver together produces a stunning contrast as the red copper bands are swirled into a silver matrix. Unfortunately, these rings are subject to corrosion between the silver and copper layers, as copper is much more active and thus more easily oxidized than the silver. The corrosion rate is intensified if the ring is worn, as salty perspiration provides a great electrolyte. Many artisans try to avoid using copper, as it readily corrodes when it comes into contact with a less active metal. By using noble metals such as gold, titanium, palladium and others, bimetallic corrosion can be stymied.

Bimetallic corrosion does occur in coins, but unless you're swimming in salt water for a few weeks with a pocketful of change, you probably won't notice it. I recently purchased a handful of inexpensive coins from a local coin dealer, sifting through a large bin. Most of them were in poor shape, having been done no favors by all of the sweaty hands covering them in electrolyte.

The more valuable coins, are, of course, neatly separated from one another in inert coin holders. There's a reason you'll never find metal coin holders in a coin shop; if you've got your coin

Figure 4.5 Wedding rings in Mokume Gane style.
© Getty Images.

collection at home in a metal coffee can, take it out immediately—you've essentially created a giant battery. If you don't believe batteries create corrosion, just examine some old batteries that were left too long in an electrical device—there's a good chance that all sorts of nasty chemicals will have leaked out. Or examine the battery terminals in your automobile—if they're encrusted with crud it is time to get out the wire brush.

4.18 BIMETALLIC CORROSION IN PLATED COINS

Bimetallic corrosion is not supposed to occur on plated or clad coins—if there is no electrolyte between the metals then electrons can't flow and corrosion can't occur. When the 1982 US copper-plated zinc penny was first proposed, there were few concerns about corrosion, even though zinc is a fairly active metal. As the zinc was to be safely ensconced inside, free from moisture and salt and oxygen, there were few concerns about it oxidizing. However, what the designers of the coin failed to realize was that by making the copper coating very thin they were making it easy to breach.

If the plating on a coin is compromised, moisture and oxygen will enter, forming an electrolyte as it gets in between the layers. If this happens your penny has become a battery, specifically a galvanic cell, with a measurable voltage. The voltage is not much, but if you connect five copper-zinc pennies in series you can actually power an LED light, the five-cent battery being quite a bargain. The voltage set up across the two metals in a copper-zinc battery can be as high as 1.1 V, an acceptable amount, given that a typical flashlight battery is 1.5 V. However, voltage is only part of the story when it comes to electricity, and in a copper-zinc penny only a minuscule amount of current flows, along the lines of 0.1 mA.

Amperage is a measure of how much current flows, while voltage refers to the potential difference between two electrodes. Voltages are determined by looking at the oxidative abilities of each metal in the two electrodes—to generate the most voltage you would want to choose two metals that differ widely in their ability to hold on to electrons. Zinc, being the more active metal, tends to give up its electrons easier than copper, therefore electrons will flow from the copper to the zinc. The higher the voltage the more current you could potentially generate.

Turning a penny into a battery might be well and good except for the fact that oxidation results in corrosion, which, as we have seen, can cause undesirable side effects. In a copper-zinc penny, terrible things can happen if the copper layer is penetrated. Most of the corrosion happens on the zinc, as it is being oxidized. The corrosion of zinc goes by the rather unappealing name of "zinc rot". In dry air zinc reacts with oxygen to form zinc oxide, and in moist air it tends to produce zinc hydroxide. The zinc hydroxide can further react with carbon dioxide in the air to produce zinc carbonate, which forms a bluish-white coating on severely oxidized zinc. As every one of these corrosion products has a density greater than that of pure zinc, as the inner zinc core corrodes it will expand, causing the penny to split open. It's no wonder they call it zinc rot.

These findings seem to indicate that the P–B ratio is not all it's cracked up to be. With a P–B ratio of 1.59, zinc's corrosion layer should provide excellent passivation. However, as the P–B ratio typically refers to oxides, it is understandable if it does not adequately predict the behavior of zinc in this situation, as other types of corrosion products predominate.

The tendency of zinc to form this nasty bluish-white corrosion layer has limited its use in coins to that of an alloy constituent or as an inner core in a plated coin. Coins made from pure zinc are not that common. A notable exception is a series of coins produced by Nazi Germany during World War II. Beginning in 1940, the 1, 5, and 10-reichspfennig coins were made from pure zinc on an emergency basis, owing to a shortage of traditional coin-making metals. In 1948 the composition would be changed to aluminum.

The most common complaint amongst collectors of these zinc coins is an unsightly white coating, which is difficult to remove. The best way to get rid of the corrosion is to dip them briefly in a strong acid. Other collectors have complained of a black discoloration on these coins, which is puzzling as most zinc compounds are lighter-colored—none are black. As these zinc coins were made as cheaply as possible, there are traces of other metals present, notably iron or lead; and oxidation of these impurities is the likely reason for the darkening. Another possibility is that the uneven surfaces of these hastily minted coins trap incoming light, causing it to appear black. As light bounces around inside the nooks and crannies it cannot escape—the absorption of this light causes the surface to appear black, for much the same reason that powdered metals appear black.

Bimetallic corrosion is often limited in coins because the alloys used are chosen for their corrosion resistance. The cupronickel alloys that make up much of today's silver-colored coinage are extremely resistant to corrosion. Furthermore, the layer of oxidation that forms on copper coins (and others) wonderfully protects them from further oxidative attack. The lack of a conductive medium is a key factor in protecting coins from corrosion—coins aren't usually exposed to the elements.

To see how coins would fare under more harsh conditions, I performed a little experiment in my back yard, placing a pile of coins in the mulch underneath some shrubbery. I wanted to see how they would fare if left outdoors, deliberately placing some of the coins on top of one another to see if bimetallic corrosion would come into play. I had all but forgotten about the coins when a few months later my wife mentioned to me that she had found a bunch of change outside in the yard, which was now sitting in a pile on the kitchen counter, scrubbed clean. The coins weren't in terrible shape, but all showed noticeable corrosion. Most had lost their luster, looking noticeably duller. The copper pennies were especially weather-beaten. The cupronickel coins fared a little better, but still showed noticeable discoloration. All of the coins were still usable, though, and considering all they had been through, fared pretty well.

4.19 NICKEL ALLERGIES

As nickel is often found in products like jewelry, watches and eyeglasses, it is often implicated as the culprit responsible for unsightly skin rashes or other allergic reactions that develop when metals come into contact with skin. However, the evidence for nickel's culpability is more than anecdotal—numerous controlled studies have shown nickel to be one of the leading causes of allergic contact dermatitis, whose symptoms include redness, itching and dryness.

When the 1 and 2-euro coins were first introduced, there were a lot of complaints about allergic reactions to the skin. Initial studies showed that these coins released over 300 times the level of nickel necessary to trigger contact dermatitis. These results were surprising, as plenty of other coins with even higher nickel concentrations didn't produce anywhere near the symptoms that these euros did. In an attempt to solve this mystery, researchers from the University of Zurich in Switzerland conducted a study wherein they attached 1 and 2-euro coins directly against the skin of test subjects for 48 and 72 hours.[11] Every volunteer showed a strong allergic reaction to the coins, as evidenced by redness and itching. The researchers also discovered that both the 1 and 2-euro coins released significantly more nickel ions than did a sample of pure nickel when immersed in artificial human sweat for 36 hours.

Next, the researchers set out to discover why the coins released such a high concentration of nickel ions. It just didn't make sense that a nickel alloy would release more nickel than the pure metal. They eventually discovered that the coin's bimetallic nature was responsible. As the coins came into contact with skin, sweat acts as an electrolyte, causing a current to flow between the dissimilar metals (cupronickel and nickel-brass) of the two parts of the coin. The galvanic potential between the two metals was measured and found to be around 40 mV, with a current flowing from the more active outer ring to the inner component. The outer ring, being oxidized, corroded five times faster than the inner part, thereby releasing more nickel ions, even though the nickel concentration was considerably lower.

4.20   SURFACE AREA AND CORROSION

There are a variety of factors that can determine how quickly a metal will corrode. When ships sink all sorts of metals end up coming into contact amongst the twisted wreckage, providing an instant undersea laboratory. In a number of studies examining the corrosion rates of metals in shipwrecks, several factors have emerged that determine the extent of galvanic corrosion that might occur between metals in contact.[12]

The characteristics of the electrolyte itself can have a big effect on the corrosion rate of metals that form a galvanic couple (the technical term for two metals in contact that undergo bimetallic corrosion). Salt water is of course an excellent electrolyte, therefore the corrosion rate in saline waters is expected to be greater than that in fresh water. The amount of dissolved oxygen in the water is also a significant factor. In waters with low oxygen concentrations, the corrosion rate is slowed. The relative surface area of each metal in contact is important. The more contact between two metals, the greater the potential for electrons to flow between them. The nature of each metal in terms of the electrochemical potential is also important—the more dissimilar the oxidative ability of the metals in contact, the more likely electrons are to flow from one metal to the next. The nature of the existing surface layers of corrosion on two metals in contact is also a factor—if one or both metals are already oxidized to some extent, then further corrosion will be slowed. Even if the conditions for corrosion of two metals in contact are not ideal, their corrosion rates are still significantly greater as opposed to a situation in which either metal is in contact with a nonmetallic substrate.

Another factor is the difference in size between the two metals in contact. Neil North of the WA Maritime Museum in Australia, in a study exploring galvanic coupling in shipwrecks, explains the effect of surface area on the corrosion rate of a brass coin in contact with a larger iron object:

> The effect of different relative surface areas can be visualized by considering the total effect to be similar for both metals. With the smaller surface area this effect is concentrated onto a small area and so produces a large result in terms of changing the corrosion rate whereas the same action spread over the larger surface area metal will only produce a small change compared to the corrosion rate in the absence of any galvanic effect. For this reason we would expect a large iron cannon to give good protection to a brass coin which was in contact with it, whilst there would be negligible increase in the corrosion attack on the iron cannon. Conversely, an iron rivet in a sheet of copper would be expected to corrode rapidly but only giving minor protection to the copper. (Excerpt reproduced from ref. 12 with permission from John Wiley and Sons, Copyright 2007.)

North also examined the size difference between two metals in contact by considering the increased resistivity that results from increased distance, drawing the following conclusions:

> In general, as the distance between any point on the surface of one metal and the nearest part of the other metal increases so does the electrical resistance of the intervening solution. As the resistance increases the magnitude of the galvanic effect decreases. In practice this leads to an uneven distribution of galvanic effect over the surface of the two metals. For example, consider a small silver coin in contact with a large sheet of copper. The silver coin will be protected by the copper and, since all parts of the small coin are relatively close to the copper sheet, protection will be relatively even over the entire surface of the coin. By contrast the copper sheet will suffer increased corrosion but since some parts of the sheet will be much further away from the silver than others, this attack will not be evenly distributed. The copper adjacent to the coin will suffer a much greater increase in corrosion than will the more remote areas. (Excerpt reproduced from ref. 12 with permission from John Wiley and Sons, © 2007 Wiley-VCH Verlag GmbH & Co. KGaA.)

One big takeaway from this research is that size matters. A copper coin that was stored in a silver box would experience significantly more corrosion than if that same copper coin were in contact with another silver coin, or in a smaller silver box. The silver box would act as an electron sink, proving to be a willing recipient for all the electrons the copper coin wants to give.

4.21 DATING WITH CORROSION

As we have seen, there are plenty of beneficial aspects to corrosion, among them toning and passivation. Another rather intriguing benefit of corrosion is its use in the dating of ancient artifacts. As there are not a lot of dates on old relics it can be difficult to determine their age with certainty. However, one electrochemical technique, known as voltammetry, holds exciting promise in dating artifacts, with one caveat—they must contain some corrosion.

The basic premise is simple: metals are good conductors, and corrosion products are not. These corrosion products offer measurably more resistance than uncorroded metals, so by comparing the amount of current that flows through a metal artifact to the amount of current that can flow through a similar metal with no corrosion, the thickness of the corrosion layer can be inferred, and with it the age of the artifact. There are some assumptions that need to be made, however, the biggest being that the rate of corrosion remains uniform over time. Even though a uniform corrosion rate is not always necessarily the case, promising results have still been achieved using this method—when artifacts of a known age were tested, the age determined was within an acceptable range.[13]

Voltammetry involves applying a variable voltage and measuring the resulting current flow (amperage). The specific method used in one study is as follows:

> To carry out the electroanalytical experiments, the researchers impregnate a graphite bar electrode with paraffin and dab the surface of the artifact with it. A few nanograms of the sample surface stick to the electrode, which is then dipped into an aqueous electrolyte. This causes almost no damage to the object. Copper oxide microparticles result in very characteristic peaks in the resulting current–voltage curves. (Excerpt reproduced from ref. 14 with permission from John Wiley and Sons, © 2014 Wiley-VCH Verlag GmbH & Co. KGaA.)

One study attempted to determine the age of a number of ancient coins and other copper-based artifacts by looking at the ratio between cuprite (Cu_2O) and tenorite (CuO) in the samples. A thin layer of Cu_2O will form first and over time upon exposure to the air it will further oxidize into CuO. To verify their hypothesis that both oxide products could be found on a single object, they analyzed a 1770 copper coin that showed red areas with black specks. A scanning electron microscopy (SEM) analysis revealed that the red areas were cuprite and the black specks were indeed tenorite, confirming their hypothesis.

After successfully dating a number of objects, they summarized their findings as follows:

> To illustrate the applicability of the proposed method... sampling was applied to a water pitcher from the Caliphal period and a Montefortino helmet of Roman age found in Valencia (Spain). Insertion of the corresponding data points into the calibration curve provides corrosion times of 1050 ± 80 and 2150 ± 150 years, respectively, in agreement with dating derived from the analysis of the archaeological context (accompanying ceramics, *etc.*).
>
> In summary, the application of the VMP (voltammetry of microparticles) technique to nanosamples extracted from copper and bronze archaeological artifacts yields specific signatures for cuprite and tenorite which can be used for dating purposes. The peak current ratio between such signals increases monotonically (in one direction only—either increasing or decreasing but not both) with the corrosion time, thereby providing a calibration curve which can be used for dating archaeological pieces submitted to moderate or light corrosion in an atmospheric environment.[14]

The fact that corrosion can be used to date objects gives us a glimpse into its nature. Though not as reliable as, say, radioactive decay, corrosion can be counted on to leave behind a record of its doings. Ever since we purified the first metal, corrosion has been at work, work that won't be finished until it has turned every last bit of metal back into its primitive state. Corrosion provides a record of the steadily progressing march of entropy, encapsulating the natural tendency of all systems toward chaos and decay. Like death and taxes, you need not question its certainty. Whereas alchemy represents man's attempts to create perfection out of chaos, corrosion does precisely the opposite. In the meantime, the best we can do is use a little alchemy ourselves to try to keep our own little corner of the universe neat and shiny.

REFERENCES

1. Coinage act, Act of Parliament, 1870, https://www.gold.org/sites/default/files/documents/1870feb10.pdf.

2. D. Glindemann, The two odors of iron when touched or pickled: (skin) carbonyl compounds and organophosphines, *Angew. Chem., Int. Ed.*, 2006, **45**(42), 7006–7009.

3. A. Sanford, Containers for protecting copper and silver coins and the like from corrosion, *U. S. Pat. Off.*, *3877571*, 1975, https://patentimages.storage.googleapis.com/d6/d8/29/bd0941c4386078/US3877571.pdf.

4. A. R. Williams, *Who buried the $10 Million in Coins Found by a California Couple—and Why?* National Geographic, https://news.nationalgeographic.com/news/2014/02/140226-gold-coins-hoard-california-discovery-numismatics/, 2014.

5. I. D. MacLeod and N. A. North, Conservation of corroded silver, *Stud. Conserv.*, 1979, **24**(4), 165.

6. R. R. Van Ryzin, Drinking and Coin Cleaning a Deadly Mix, *Numismatic News*, https://www.numismaticnews.net/flipside/drinking-and-coin-cleaning-a-deadly-mix, 2007.

7. W. W. White, Sulfide and the toning process, *Numismatist*, 1990, 33.

8. J. Porcayo-Calderon, R. A. Rodríguez-Díaz, E. Porcayo-Palafox and L. Martinez-Gomez, Corrosion performance of Cu-based coins in artificial sweat, *J. Chem.*, 2016, **2016**, 1–11.

9. N. B. Pilling and R. E. Bedworth, Mechanism of metallic oxidation at high temperatures, *Chem. Metall. Eng.*, 1922, **27**, 72.

10. L. D. Rosenhein, The household chemistry of cleaning pennies, *J. Chem. Educ.*, 2001, **78**(4), 513–515.

11. F. O. Nestle, H. Speidel and M. O. Speidel, High nickel release from 1- and 2-euro coins, *Nature*, 2002, **419**, 6903.

12. N. North, The role of galvanic couples in the corrosion of shipwreck metals, *Int. J. Naut. Archaeol.*, 1984, **13**(2), 133–136.

13. Angewandte Chemie International Edition, *Dating by Electrode*, https://www.chemistryviews.org/details/ezine/6399591/Dating_by_Electrode.html, 2014.

14. A. Doménech-Carbó, M. T. Doménech-Carbó, S. Capelo, T. Pasíes and I. Martínez-Lázaro, Dating archaeological copper/bronze artifacts by using the voltammetry of microparticles, *Angew. Chem.*, 2014, **126**(35), 9416–9420.

CHAPTER 5

Paper Money

> *"Paper money eventually returns to its*
> *intrinsic value—nothing."*
>
> *Voltaire*

If someone asked how much money you had, it is doubtful you would count the number of coins in your pocket, or even the number of coins stashed away at home. If you liquidated your bank account and received your life savings in bags of coins, you would probably be annoyed—a suitcase full of paper money would be more like it. Although "cold hard cash" sounds like an apt metaphor for coinage, it usually refers to anything but. In the not-too-distant past, coins reigned supreme, but nowadays they have been relegated to pocket change. Although not quite obsolete, they have lost their relevance. In the not-too-distant future paper money might well meet the same fate, but for now, paper (to use the term very loosely) money reigns supreme.

5.1   THE ORIGIN OF PAPER MONEY

Paper money has a completely different trajectory to that of metallic money. As far as we know, the first paper money originated in the Tang Dynasty in China, in the 7th century AD. These very first banknotes were handwritten on one side only. Two-sided printed banknotes did not arrive until much later.

The Chemistry of Money
By Brian Rohrig
© Brian Rohrig 2021
Published by the Royal Society of Chemistry, www.rsc.org

As the Chinese invented paper hundreds of years earlier, it is only fitting that they would also invent paper money.

The word paper is derived from the word *papyrus*, which refers to the material from which ancient scrolls were comprised. The pith—the innermost part of the stem—would be extracted from the papyrus plant and cut into thin strips that would be joined together to make scrolls. Paper, on the other hand, involves a chemical process in which wood is changed into paper. The making of papyrus is very much a physical process, akin to taking a piece of wood and cutting it into thin strips, which is not at all the same as making paper.

The earliest paper was made from a mixture of mulberry bark and rags of cotton and hemp, which were added to water and then literally beaten to a pulp. This pulp was then spread flat and allowed to dry. The end result was paper. This early paper was still rather crude and not all that strong, primarily because most of the lignin remained.

Lignin is a component of the cell wall that lends rigidity to plants. It is the glue that holds the various components of the cell wall together, creating the stiffness that allows trees to grow to much greater heights than herbaceous plants, which contain less lignin. One common variety of lignin has the formula $C_9H_{10}O_2$; as its molecules do not have the $2:1$ hydrogen to oxygen ratio it is not classified as a carbohydrate.

Lignin is very resistant to rot—a piece of old wood on the forest floor is mostly lignin, which is the last part of wood to decompose. Counterintuitively, the presence of lignin actually makes paper weaker, not stronger. As papermaking advanced, more creative ways were discovered to remove lignin from pulp, usually by using strong alkaline compounds, leaving behind mostly cellulose, which made for a much stronger paper.

Like the rest the world, the Chinese would eventually utilize metal coins for money. One common type of coin was made from copper, with a hole in the middle, making it easy to tie a string through multiple coins for ease of transport. As merchants tired of carrying around so many heavy coins, eventually someone came up with the idea of leaving one's stash of coins with a trusted intermediary, in return receiving a piece of paper that served as a promissory note, which could be redeemed at any time. In the 12th century AD, during the Song Dynasty, the government would officially issue paper money, owing to a copper shortage.

Paper money was introduced to Europe through explorers to the East who reported back on this strange practice. One such voyager was the Venetian explorer Marco Polo, who spent some time with Kublai Khan, emperor of the Mongol Empire during the latter half of the 13th century. Polo found the idea of paper money intriguing, penning an entire chapter on money in his seminal work *The Travels of Marco Polo*, entitled "How the Great Khan Causes the Bark of Trees, Made into Something Like Paper, to Pass for Money All Over His Country".

Other explorers to China reported back with similar stories of paper money, but still, the practice didn't catch on in Europe until 1661, when the Bank of Stockholm in Sweden issued paper banknotes during a silver shortage. They were a welcome replacement for the copper plate money that was prevalent at the time. Even though technically considered coins, they were rectangular in shape. The heaviest one had a value of 10 dalers (the precursor to today's dollar) and with dimensions of 30 cm by 70 cm weighed in at nearly 20 kg! They were the largest coins ever made, being stamped in each corner and in the middle with the denomination, making each plate look like five coins embedded in a large sheet of copper (Figure 5.1). As copper depreciated, the plates grew in size accordingly. As no one enjoyed the idea of lugging around such cumbersome "coins," when given the opportunity to exchange them for paper money, most people jumped at the chance.

Figure 5.1 Swedish daler.
© Stack's Bowers Galleries.

It didn't take long before other countries followed Sweden's lead, with private banks often printing their own money. The first paper money in America was issued in 1690 by the Massachusetts Bay Colony as a way to pay soldiers who had fought the French in Canada. The Bank of Scotland issued its first banknote in 1695. The first paper francs came out in 1795. Mexico began issuing peso banknotes in 1823. Belgium came out with banknotes in 1837. The Japanese yen was first introduced in 1871, with the first paper yen debuting the following year. It took Germany until 1873 to get on board with paper, with the introduction of the gold mark. Even as they spread across the world, banknotes were slow to be accepted—silver and gold coins were still the preferred method of currency.

Banknotes have been printed on a lot more than paper. Leather, silk, sealskin and even playing cards have all been used as the substrate for money. Paper would eventually give way to a much more durable cotton-linen fabric, which to this day is still the material of choice for much of the world's banknotes.

5.2   COTTON-LINEN BANKNOTES

In the US, banknotes are made from a specially formulated cotton-linen blend furnished exclusively by the Crane Company. The Cranes have been supplying currency paper to America since its inception. The founder of the Crane Company, Thomas Crane, sold the paper to Paul Revere to make banknotes for the colonies in 1775, shortly before declaring independence from Great Britain. The Crane Company has made life difficult for counterfeiters, incorporating any number of innovations into their paper. In 1928, they began randomly distributing red and blue silk threads throughout banknote paper. If you examine any US banknote under a microscope, you can see that the threads are still being used.

Today the Crane Company supplies currency paper for about 50 other countries in addition to the US, and even prints currency for a number of those countries. The composition has changed over the years. During World War II a shortage of linen forced the company to change its formulation from 75% linen to 50% linen. In 1956, the composition changed to its current ratio of 75/25 cotton-linen, the change being made to accommodate newer printing techniques which enabled the printing of money on dry paper.

Linen is one of the toughest fibers known, adding strength and durability to paper. If you want a really tough, high-quality paper you'll go with one that contains linen. Linen adds body without adding bulk.

A big difference between high-end resume quality paper and the cheap stuff is stiffness. Stiffness is a function of strength, with the strength of a fabric being directly proportional to its fiber length, among other things. Lengths up to a meter are not uncommon with linen fibers, with 20–40 cm being the norm; cotton fibers are only around 2–3 cm long. This added length gives linen fibers considerably more tensile strength than cotton; an added benefit is that linen's tensile strength increases by about 20% when wet. Tensile strength is a measure of how much tensile stress (stretching) a fiber can withstand before it snaps in two.

Linen has traditionally been valued for its alleged anti-microbial properties as well as its resistance to rot and mildew; controlled studies, however, have failed to establish any anti-microbial effect for linen. It is most likely that linen's great strength and durability alone has enabled it to endure the test of time.

5.3 THE TOUCH, THE FEEL OF COTTON

If you were blindfolded and handed a sheet of linen paper and one made from 100% cotton, you could easily differentiate between the two—cotton paper is as soft as a baby's bottom. Some banknotes, such as the euro, eschew linen altogether and go with 100% cotton, making for a flimsier, less rigid note. This lack of linen has had a noticeable effect on the euro's lifespan, which is only half that of a typical cotton-linen banknote. To counter this problem, all euro banknotes are now treated with a protective varnish in an attempt to make them more durable.

If you visit an art supply store you can purchase cotton rag paper. As its name implies, the cotton to make this paper comes from rags, not the soiled rags you might find under your kitchen sink, but rather unused scraps from the textile industry. These rags are already processed and can be purchased in bulk at low cost.

Cotton rag paper can easily be made in the kitchen. The first step is to take a 100% cotton tee shirt and cut it into tiny pieces. Once the pieces are softened by soaking overnight in water, they can be put in a blender until reduced to a pulp. The pulp can then poured onto a screen of some sort—an old window screen works fine—and flattened out. When the pulp dries, you will have a piece of cotton rag paper.

Cotton rag paper is very durable and speaks of quality and refinement. The finest papers are made from cotton rag—even the name evokes a certain chic euro trashiness. Cotton paper tends to resist coloring and deterioration, lasting centuries longer than paper made from wood. Important documents should always be printed on cotton paper—you don't want your college diploma or marriage license someday crumbling into dust. Most quality cotton rag paper has a watermark which discloses the percentage of cotton, verifying its lineage.

For many years, scraps of undyed denim left over from making jeans provided a rich source of rags for paper mills. However, the rise of stretch fabrics like skinny jeans and the like complicated matters for the Crane Company and other makers of currency paper. Stretch fabrics often contain Spandex, which is all but impossible to remove from the fabric, and when used in paper compromises the strength of the final product. Therefore, to ensure the quality of their stock, most makers of currency paper today get their cotton directly from cotton mills, who guarantee it to be free of additives.

5.4   THE FIBROUS NATURE OF PAPER AND BANKNOTES

If you look at a cotton-linen banknote under the microscope, and then look at an ordinary piece of paper, they will look surprisingly similar. You will see a lot of overlapping strands on top of still more strands. Both are made from plants, which have been crushed and pulverized and treated until the desired properties are attained. Cotton, of course, is picked from a cotton plant, whereas linen is made from flax. The seeds of the flax plant are called linseeds (and sometimes the flax plant itself is called linseed), and are used to make linseed oil.

Linseed oil is quite versatile—its uses range from paint thinner to wood finishes to health food supplements. Oxidized linseed oil is a key component of linoleum, the floor covering. Not to be outdone, cottonseed oil is just as versatile—you can use it to deep-fry junk food, or if you prefer, make soap. It is also an excellent insecticide.

When magnified, linen fibers resemble a stalk of bamboo—they are firm and rigid. Cotton fibers are wispier and are often twisted, vaguely resembling a fragment of DNA. Both types of fiber are clearly visible under the microscope.

The fibers in paper are similar to those in a banknote, but differ widely depending on what types of plants are used to make the paper. Trees with softer wood are preferred for papermaking as they are easier to blend into pulp, with coniferous trees such as pine and spruce topping the list. If hardwoods are used, they are usually the softer hardwoods like aspen, which take less energy to pulverize than harder woods—you won't see much paper made from ironwood.

You can actually make paper from just about any plant stem, herbaceous or woody, depending on how strong you want the paper to be. Among the weakest paper is newsprint, which is mostly made from recycled paper. Anytime paper is recycled the cellulose fibers get shorter, which is why you can only recycle a particular piece of paper about five times. Paper can't be made from 100% recycled products; it typically needs to be supplemented with some virgin wood pulp to increase the fiber length. If you look at a sheet of newspaper under the microscope you can see that its fibers are noticeably shorter than ordinary paper, as well as being considerably less densely matted.

5.5 CELLULOSE

Cotton and linen are both made of about 90% cellulose, which is the primary component of plant cell walls. Cellulose is the most abundant plant product on earth—plants make over 10 billion tons of it every year. It is indigestible to humans, and even animals that break down cellulose can't do it on their own. It doesn't matter how long cows chew their cud, without the action of specialized bacteria in their rumen, or first stomach, they cannot digest cellulose. The same goes for termites—they are also reliant on bacteria to break down the cellulose in wood.

The number of products made from cellulose is surprising. Celluloid (as in old movie film) and cellophane (as in tape) are both made from cellulose, albeit with a bit of chemical tweaking. Ping-pong balls are made from celluloid, and like the old movie films, burn very well. Fabrics like rayon and acetate are made from cellulose, and even though cellulose is a natural fiber it might be a stretch to call these fabrics natural—the cellulose is subjected to a lot of chemical manipulation before it emerges as a regenerated fiber.

Polymer banknotes have recently been introduced to great fanfare in a number of countries, including Canada, Australia, and the UK. However, the banknotes they replaced were also made of polymers, so it should really be called plastic money. Cellulose is just as polymeric as plastic, both being made of repeating subunits, or monomers; with cellulose the monomers are glucose. These glucose monomers are not the typical sugar molecules $(C_6H_{12}O_6)$ that flow through your blood and give you energy, however, but rather are glucose molecules that are missing an H_2O—think of them as dehydrated glucose, giving cellulose the rather odd-looking formula of $(C_6H_{10}O_5)_n$, with n representing the number of these repeating glucose subunits, which can number from the hundreds into the thousands, with around 10 000 representing the upper limit. Cellulose is an example of a condensation polymer, as a water molecule is removed (condensed) every time two normal glucose molecules come together.

Cellulose is notoriously insoluble, which explains why your banknotes don't fall apart when they get wet. Even though cellulose doesn't break down in water its surface can adsorb a fair amount of it. Cotton and linen fabrics, being primarily composed of cellulose, are weakened by water but don't break apart when they get wet; otherwise we would need to dry clean all of our clothes. Wood, and other materials containing cellulose, can also get wet—the water doesn't penetrate very far, though. I still recall an overnight camping trip as a boy; it had been raining heavily and we wanted to make a fire, but there was no dry wood anywhere to be found. One of my cohorts had the wherewithal to cut open a wet log and extract kindling from the inside, which was bone dry. In no time we had a roaring fire.

Each cellulose chain contains a lot of hydroxyl (–OH) groups. The oxygen of one hydroxyl group bonds with the hydrogen on a hydroxyl group of an adjacent cellulose chain. These chains form a dense sheet, which in turn bonds to other sheets of cellulose. Hydrogen bonding not only keeps the cellulose strands connected within a single sheet, but also serve to bond one sheet to another, forming an impenetrable barrier to water. Water can seep in and break a few hydrogen bonds between the chains and collect on the surface, but the cellulose chains themselves are not affected.

The primary component of actual paper is also cellulose, which begs the question as to why when paper gets wet it turns into a sodden mess, unlike most other cellulose products. The cellulose in paper goes through a lot more distress than that in fabrics. In the pulping process a lot of the glucose dissolves in the water, and when this glucose dries it acts as a bonding agent between the cellulose chains. When it gets wet the soluble glucose dissolves and the paper falls apart.

5.6 STARCH

Another key component of paper that contributes to its breakdown in water is starch. A piece of paper can contain anywhere from 5–10% starch. If you starch your shirt it will be stiff; to make gravy nice and thick you need starch. In the same way starch adds body and thickness to paper, making it tough and durable. The word *starch* derives from the old German word *stearc*, which means stiff or strong. Unlike cellulose, starch turns into mush when it gets wet, readily forming hydrogen bonds with water. As a result, starch-based papers break down quite easily when they get wet. Like cellulose, starch is a polymer composed of units of glucose (minus a water molecule) that are bonded together in either a long unbranched chain (amylose) or a branched chain (amylopectin) (Figure 5.2). Both forms have the same formula as cellulose—$(C_6H_{10}O_5)_n$.

There is a huge difference though, between starch and cellulose. The glucose units in cellulose are bonded together much more strongly than they are in starch—these strong bonds are what make cellulose almost indigestible (Figure 5.3). This difference in bonding arises from subtle differences in the glucose

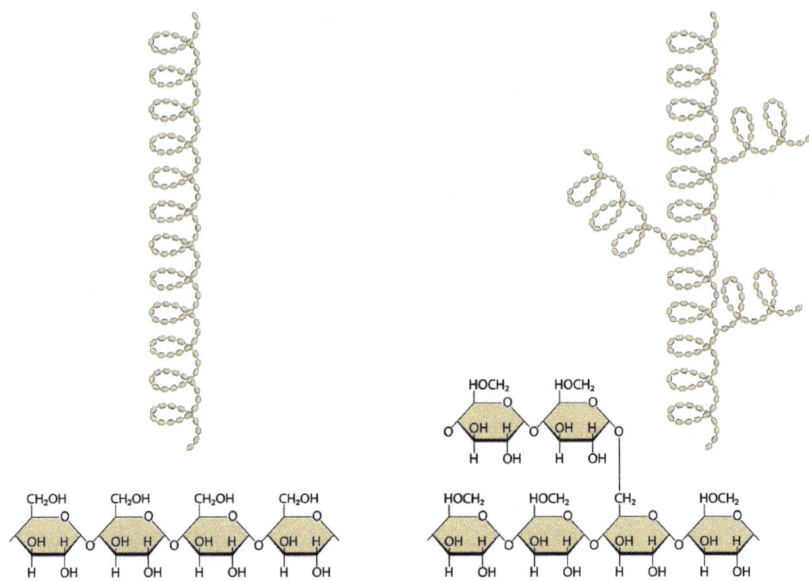

Figure 5.2   Differences between amylose and amylopectin.
© Shutterstock.

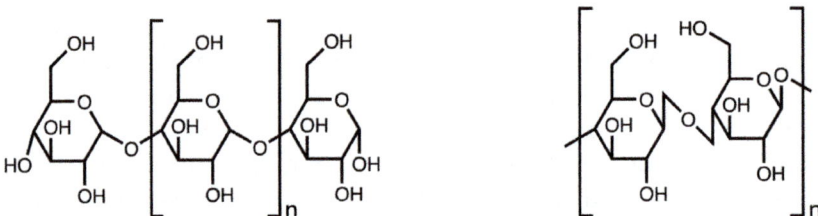

Figure 5.3   Bonding in amylose (left) and cellulose (right).

molecules that make up the polymer. Starch is made up of
α-D-glucose while cellulose is made up of β-D-glucose. In α-D-
glucose, opposite ends of the glucose molecule contain a
hydrogen atom and a hydroxyl group in the exact same position.
If you examine a model of this molecule you can see that on each
end there is a hydrogen atom on "top" and a hydroxyl group on
the "bottom". The bonds holding these glucose units together
are known as α-1-4 glycosidic linkages. The numbers refer to the
specific carbons within the glucose—the number 1 carbon of one
molecule bonds to the number 4 carbon of an adjacent molecule.

(A glycoside is the technical term for a glucose molecule that is lacking a water molecule. Think of a glycoside as a sugar that will yield glucose after hydrolysis.)

The type of glucose that makes up cellulose, on the other hand, is not as symmetrical as the glucose molecules that make up starch. If you were to examine a model of β-D-glucose, you would see that on one end of the molecule there is a hydrogen on top and a hydroxyl group on the bottom, but on the other side the positions are reversed—the hydrogen is on the bottom and the hydroxyl is on top. When two β-D-glucose molecules link up they always do so by joining together opposite ends, therefore the side with the H "on top" will bond to the side of an adjacent glucose molecule with the H "on the bottom". The bonds holding these glucose units together are known as β-1-4 glycosidic linkages. Therefore, the glucose units that make up starch will all be arranged in a nice symmetrical row, with each glucose oriented the same way. With cellulose, however, it would be like every other glucose unit is upside down.

The way the glucose units bond in cellulose serves to make the chain much straighter than it is in starch. As a result, the chains can pack together tightly, with a great deal of hydrogen bonding between the chains. This densely packed structure is very tough, making it difficult for water to penetrate, which makes sense as cellulose gives structural support to plants. The bonding in starch tends to make the chains zig zag, leading to a structure that is almost helical. The β-1-4 glycosidic bonds in cellulose end up being much stronger than the α-1-4 glycosidic linkages in starch—primarily owing to bond angles that result in a more stable configuration, helping to explain why cellulose is so much more difficult to digest than starch.

5.7   STARCH TESTING

As real money does not contain starch, a commonly used method to detect counterfeit bills involves testing for the presence of starch. The easiest way to test for starch is to apply iodine, usually in the form of an iodine-potassium iodide solution. Tincture of iodine (which contains alcohol) works well, too. In these solutions, iodine exists as the triiodide ion (I_3^-), which forms when diatomic iodine reacts with the iodide ion.

A solution of iodine will be yellowish in color, but in the presence of starch it will turn a deep purple, almost black, color.

Iodine and starch bond together to form what is commonly referred to as the starch–iodine complex. Only the linear form of starch (amylose) forms the starch–iodine complex. The I_3^- ions are just big enough to squeeze into the spaces within the somewhat helical-shaped starch polymer chain. A negative ion is always smaller than its elemental form, as the loss of electrons enables the now overpowering nucleus to shrink the electron cloud considerably. If elemental iodine crystals are added to starch nothing happens, nor does anything happen if the iodide ion alone is added to starch, such as you might find in a solution of KI. It is necessary to have both elemental iodine as well as iodide ions so that I_3^- can form.

In countries in which cotton or cotton-linen banknotes are used, it is common practice for a cashier to mark the banknotes that you present as payment with a special anti-counterfeiting pen that writes with yellow "ink". Cashiers usually won't bother marking $1-dollar bills, but typically will mark anything above that. A $50 or $100-dollar bill will almost always be marked. These pens are commonly available in office supply stores, and are a wonderful example of chemistry in action. The "ink" in these pens is nothing more than a solution containing the triiodide ion.

The absence of a black mark does not mean that the bill is genuine, however—it only means that it doesn't contain starch. Although the iodine pens do work as advertised, they provide a false sense of security, as there are plenty of types of paper that do not contain starch. If you do a quick web search for "starch-free paper" you will find a plethora of sites offering 75% cotton–25% linen starch free paper. Many companies even advertise their paper as "Banknote starch free paper".

There are plenty of types of paper floating around the house or lab that don't contain starch. Testing for it only requires a little bottle of tincture of iodine, readily obtainable from your local pharmacy. Another way to "test" for starch, without using iodine, is to see if the paper breaks down in water. If it falls apart in water it probably (but not always) contains starch. You can then verify your hunch by adding a drop of iodine. Coffee filters, as well as laboratory filter paper, are starch free. The same goes for paper towels and facial tissues—they are designed to get wet and not fall apart. Newsprint doesn't contain starch either, which is why if you spill your coffee while reading the paper you will have something handy with which to sop it up.

A great way to demonstrate the formation of the starch–iodine complex on a large scale is to spray some starch from an aerosol can (used for ironing) into a large beaker of water. If some iodine solution is added it will turn it a deep purple. Adding some vitamin C to this mixture will turn it colorless, the vitamin C reduces the I_2 into colorless I^-. If you want to bring back the color, you can add some bleach, which oxidizes the I^- back into I_2. Therefore, if you want to foil the iodine marking pens all you have to do is treat your starchy counterfeit bills with vitamin C.

5.8 "BURNING" MONEY

An interesting way to demonstrate the durability of these cotton-linen banknotes involves dipping a banknote in a 50:50 solution of isopropyl alcohol and water and then setting it on fire. Even though it will be engulfed in flames, the banknote itself will not be harmed. I had a friend who carried in his pocket half of a \$50 bill as testament to the fact that this experiment can go horribly awry if the correct concentrations of alcohol and water are not used, so it's best to practice on a lower denomination bill first.

The water is key to the success of this demonstration. Only the alcohol burns, not the water. (Incidentally, any alcohol concentration below 50% will typically not burn, hence the designation of 100-proof liquor, which equates to 50% ethyl alcohol, which at one time was established as the minimum legal allowable concentration of alcohol that could be sold—if the liquor burned, it was "proof" that it was of sufficient concentration.) As the water is heated, its temperature is raised and it vaporizes. Water's rather high specific heat of 4.18 $J g^{-1}$ °C, along with its high heat of vaporization of 2260 $J g^{-1}$, makes it perfectly suited for absorbing large quantities of heat. The water absorbs so much heat that it prevents the banknote from ever reaching its kindling temperature.

5.9 WRINKLED MONEY

As they are made of cotton and linen, banknotes have a lot more in common with your clothes than you might think. If you don't take a load out of the dryer fast enough, your clothes will be reduced to a bundle of wrinkles, necessitating a lot of unpleasant ironing. When clothes are washed, hydrogen bonds are broken between the cellulose molecules. Every serious runner knows not to wear a cotton shirt—runners even have an acronym to express their disdain for cotton, expressed as the ABC of running (anything but cotton).

Cotton fabrics hold water, becoming wet and heavy and uncomfortable. The newer synthetic fabrics wick moisture away from your body through capillary action, keeping your clothes relatively dry. These synthetic fabrics are also more naturally water repellant—less moisture penetration means less hydrogen bond breaking; if fewer hydrogen bonds break then fewer new hydrogen bonds will form, resulting in fewer wrinkles.

You can easily demonstrate hydrogen bonding in a cotton-linen banknote by getting it wet and letting it lie flat—when it dries it will be all wrinkly. Paper gets even more wrinkled than fabric if it gets wet; if you get a book wet it is ruined—an overwhelming number of original hydrogen bonds are broken and then all sorts of new ones haphazardly form as the paper dries. You get bed-head in the morning from lying on your hair all night—hydrogen bonds are broken by the weight of your head putting pressure on your hair, causing the hair to become matted and folded at odd angles. If you go to bed with wet hair the bed-head is much worse—the water serves to break even more hydrogen bonds.

However, water is only half the equation when it comes to wrinkling. The other half is heat. There is a scientific reason why some garments should only be washed in cold water and never put in the dryer. Every amorphous polymer has what is known as a glass transition temperature, which is the temperature at which it transitions from a hard and brittle glass-like state into something softer and more malleable. Like glass, an amorphous solid is one with no discernible crystal structure; instead of its molecules being arranged in a nice orderly fashion they instead appear to be randomly tossed about, even though they are locked in place.

If a polymer is heated above its glass transition temperature it will soften and become pliable; it does remain a solid though, as it is still below its melting point. If a soft polymer is cooled below its glass transition temperature it will become hard and brittle, which is why a racquetball immersed in liquid nitrogen will shatter when dropped.

Fabrics with a relatively low glass transition temperature are most easily damaged by heat—their hydrogen bonds are broken quite easily. Water can, however, effectively reduce the glass transition temperature of most fabrics. Even though cotton has a relatively high glass transition temperature of 230 °C, when saturated with water it reduces significantly, actually going down to around room temperature, which explains why a cotton T-shirt left in the clothes basket becomes so wrinkly.

When clothes are ironed the heat takes the fabric through its glass transition temperature—polymers begin to soften as hydrogen bonds break. The weight of the iron and the pressure exerted by the one who wields it forces new hydrogen bonds to form between the cellulose molecules, forcing them into a parallel alignment that will hopefully cause the fabric to be free of wrinkles. If the iron is running low on water, it will be difficult to do your ironing—but once water is added and it turns to steam wrinkles can be removed with a lot less effort.

Water helps breaks the hydrogen bonds holding the wrinkles in place, reducing the glass transition temperature.

Reading the settings on an iron is a good way to receive a quick tutorial in glass transition temperatures—on my iron, the heat settings are in the following order, from highest to lowest: linen, cotton, wool, silk and synthetic. These settings roughly correspond to the glass transition temperatures of various fabrics, with the synthetic fabrics having much lower values than the natural fibers.

The ideal fabric will possess excellent wrinkle recovery (yes, wrinkle recovery is a real thing and it is quantifiable). Fabrics with excellent wrinkle recovery can be subjected to stress for a relatively long period of time and then bounce back with little evidence of ever having been tampered with. Paper has terrible wrinkle recovery—if you fold it once the crease is permanent. If you fold a dollar bill the crease is much less noticeable—it has far better wrinkle recovery than a piece of paper. According to the US Bureau of Printing and Engraving, a dollar bill will only tear after 4000 folds.

I was curious as to how money compared with paper, so tried seeing how many times I could fold a piece of copy paper until it would tear. It took 129 folds before I saw the first tear, with 191 folds required for complete separation. I was surprised that the resulting borders between the two severed halves were not smoother; the borders were almost serrated. I didn't have the patience to test the dollar bill.

As a dollar bill is made of fabric and not paper, it stands to reason that like any fabric, it can be ironed if it gets wrinkled. But before seeing if a banknote could be ironed, I first wanted to see if a piece of paper could be ironed, as they are both made of cellulose. An iron is not supposed to get hot enough to burn paper. On the highest setting, which is for linen, an iron reaches a temperature of 230 °C, whereas the kindling temperature of paper is 233 °C. This slim three-degree margin didn't give me a great deal of comfort, however, so I set the iron to low.

The iron didn't touch the wrinkles at this setting, however, so I gradually turned up the heat, and despite my fears that the paper might burst into flames, kept increasing the setting until I got to linen, dangerously close to paper's kindling temperature. The paper flattened out, developing a textured surface, but the creases didn't budge, if anything they were more firmly impressed into the paper. I remembered to turn on the steam, which helped a little, but not much. Next, I pressed the little button on the iron that squirts out water, almost saturating it. Finally, I was making progress; some of the creases did come out, but as soon as the heat of the iron evaporated the water the paper began to wrinkle again. The paper ended up being a little smoother than before, but for the most part the creases were not at all amenable to being removed. Fortunately, there was no charring.

With the trial run of the paper completed, I now turned my attention to a dollar bill that I had wadded up into a ball. I hadn't expected to be successful in ironing the piece of paper, because it was, after all, a piece of paper, but I had high hopes for the banknote. Banknotes are made of fabric, and fabric should be ironable. I immediately set the iron to "cotton" and went at it. I tried the steam and then the water treatment, saturating the bill. Even with the iron turned up all the way, to the linen setting, I was only moderately more successful in removing wrinkles than I was with the piece of paper.

Being displeased with the results of the ironing experiment, I resorted to another method to get rid of the wrinkles. One way to dewrinkle fabrics is to run them through the washer and dryer, so I decided to send some money through the entire laundering cycle. I wadded up ten one-dollar bills, unfolded them and put them in the washer. I set the water temperature to hot added detergent and set the dial to "delicates".

With the wash cycle completed, I opened the lid to see the banknotes neatly plastered against the sides of the washer, having been subjected to centripetal forces. It looked as though I was engaged in some sort of sophisticated money laundering operation. Each of the bills was sparkling clean—probably clean enough to eat from. They were mostly wrinkle-free—the only wrinkles present were those that developed as a result of their orientation after going through the spin cycle; some of the notes were folded in half or otherwise crumpled, but even on these the wrinkles were not noticeable when straightened. It seemed the hydrogen bonds were no match for the rigors of a washing machine.

Next, I put the bills in the dryer. I didn't use a dryer sheet, as I was curious as to whether or not they would develop any static cling. After ten minutes the bills were dry. The banknotes were scattered about the dryer, some being stuck in various crevasses. I had to reevaluate my earlier assessment of the wrinkle-removing ability of the wash cycle. The bills were full of wrinkles—they may have even developed a few new ones. There was no observable static cling.

Still not satisfied with my attempts to remove the wrinkles, I resorted to other methods. I thought perhaps more heat was needed, so I soaked a dollar bill in water and microwaved it for two minutes—the bill was dry but the wrinkles were untouched. Perhaps more time was needed, so I immersed the same bill in a pot of boiling water for 30 minutes. Upon removing it, it didn't look wrinkled, but rather than let it dry on its own, on a whim I placed it beneath a stack of books, completely ruining my controlled experiment. I did get another banknote wet, though, and placed it under some books as well. Upon retrieving the boiled banknote the next day, it appeared to be wrinkle-free, finally. The bill that had been saturated but not heated was noticeably more wrinkled.

Evidently the wrinkling, and the subsequent curling of the bill, occur upon drying. Therefore, it appears that the formula for removing wrinkles is: water + heat + time + pressure + more time. Wrinkling is not just an issue for consumers, but can also present problems during the production process. In 2011, the release of a new American $100 bill had to be postponed because the newer security features were causing unwanted wrinkles in the bills.

5.10 SHRINKING MONEY

Although my dollar bills did not shrink in the wash, there is another way to shrink money, literally. If a banknote is immersed in liquid ammonia and then set out to dry, it will shrink considerably once the ammonia evaporates. Liquid ammonia is not the same thing as concentrated ammonium hydroxide, which will not work for this experiment—liquid ammonia is made by compressing anhydrous ammonia gas (NH_3); it is quite chilly, boiling at −33 °C.

If the immersion of a banknote in liquid ammonia is repeated several times, each time the bill shrinks even more. The chemistry as to why the bill shrinks is still unknown, but is most likely related to the disruption of hydrogen bonds in the cellulose by the ammonia, causing the fibers to relax and shorten. Wood placed in anhydrous ammonia will also soften, indicating disruption of the hydrogen bonding between the cellulose and lignin fibers, enabling the chains to slide past one another, making them soft and pliable.

In one notable experiment using this process, the dimensions of a US one-dollar banknote were reduced considerably—from the normal size of 15.7×6.6 cm to the much smaller 9.8×4.8 cm.[1] The bill did not shrink proportionally, however, as its length shrank by 38% but its width by only 27%. The mass, unsurprisingly, was not affected. The disproportionate shrinkage of the bill lengthwise is likely a function of how the banknote was made.

If you attempt to rip a piece of paper, it will usually rip fairly straight and cleanly in one direction, but in the other direction it will be jagged and uneven. The cheaper the quality of the paper, the greater the disparity between the way it tears lengthwise as opposed to widthwise. During the paper-making process, as the pulp is pressed and makes its way down the conveyer belt, a grain is established in the paper along the direction of the belt's travel. If you rip a piece of paper in the same direction as the grain, the tear will be more or less straight, but if you try to rip it against the grain, the tear will be all over the place—the paper is fighting you, trying to tear along the direction of the grain so as to follow the path of least resistance.

Although I'm not in the habit of mutilating money, for the sake of science I was willing to make an exception. I ripped several banknotes in half, and on every one, the tear was much more even and clean when ripped widthwise as opposed to lengthwise. On normal paper the reverse is true—it will be easier to get a clean tear if ripping it lengthwise. However, a single bill only represents a portion of a larger sheet of bills that have come off the press. If you watch money being printed as it moves through the printing press, the sheets of bills begin as longer rolls, with the paper moving in the direction of the grain. Therefore, rather than banknotes rolling off the press lengthwise, they actually roll off widthwise before being cut into individual bills. As the cotton-linen fabric that banknotes are made from has a grain, the fibers are aligned differently in one direction than in the other, resulting in the banknotes shrinking more one way than the other.

5.11 PATHOGENIC MONEY

Despite all the abuse it goes through, for the most part paper money holds up pretty well. One analysis revealed that a £20 note will pass through 2328 different hands throughout the course of its lifetime.[2] There is probably nothing else that we possess that gets passed around as much as money; the ability of money to handle all of this handling testifies to its durability, but on the other hand makes it a germaphobe's worst nightmare. Unlike metallic money, which possesses some antimicrobial properties (especially copper), paper money has no such ability. No respectable restaurant will have its workers handle food after taking your money without having them first wash their hands or don gloves. My mother gets especially upset if she views this practice, and will always inform the manager if she notices it.

Numerous studies have been undertaken in an attempt to discover if money really is as dirty as people think it is. One study tested 1280 banknotes from food outlets in ten different countries, in the hopes of using this data to promote good hygiene amongst food handlers.[3] The bacteria per cm^2 were counted on a variety of banknotes. Even though there was a lot of variation from bill to bill within a given country, every single banknote was found to be crawling with bacteria, and not just a little bit either. It appears my mother's concerns were warranted after all—money is indeed quite disgusting.

The researchers also discovered that banknotes from less economically prosperous countries had larger bacterial counts than those from more prosperous countries. Cotton-linen banknotes from China, Nigeria and Mexico had larger bacterial counts than those from Ireland, the US and the UK. The researchers attributed these findings to differences in sanitation.

The type of material from which the banknote was made played a significant role in its bacterial count. Banknotes made from cotton-linen had a significantly higher bacterial count than those made from plastic. In currency tested from Mexico, which uses both types, the bacterial count on the cotton-linen banknotes was four times as great. The age of the banknote also played a role—older banknotes were found to have higher bacterial counts than newer ones.

In another study, researchers set out to determine how well banknotes could transmit pathogenic bacteria.[4] Their first step was to inoculate some broth with a variety of nasty microorganisms, including *S. aureus* (MRSA), *Enterococcus feaecium* and *E. coli*. After letting this broth stew for 24 hours, it was spread onto the surface of various banknotes. The bacteria were then counted at various intervals, both before and after drying.

Bacteria thrived the best on the Romanian leu, with active cultures from all three strains present even after six hours of drying; one type was present even after a day of drying. The euro, US and Canadian dollars, and Indian rupee all had various levels of contamination. Only the Croatian kuna showed no bacteria after three hours, the shortest time interval recorded.

The researchers then tested whether or not these bacteria could be transmitted through human contact. They tested bills from three countries. After contaminating the bills, they allowed them to dry for 30 minutes. Three subjects then rubbed them in their hands for 30 seconds and then touched some agar. After

24 hours the bacteria were counted. No bacteria were transmitted from the euro, but all three subjects successfully transmitted bacteria from the US dollar and the Romanian leu, with the leu having the highest transmission rate. Of all the currencies tested, only the Romanian leu was made from plastic.

In yet another study, one-dollar bills collected in Ohio, USA found that 94% of the bills were contaminated with potentially pathogenic bacteria.[5] Another study of $1 banknotes in New York was able to isolate almost 400 different bacterial species from various bills![6]

The fibrous surface of a cotton-linen banknote provides numerous nooks and crannies that harbor bacteria; a plastic note has a much smoother surface. Although our money can harvest any number of nasty microbes, there is currently little evidence to indicate that the amounts carried on currency can actually make you sick. There is no use taking chances, though, so wash up after handling money—you never know where it's been.

5.12   DRUG MONEY

There just might be a lot more lurking on our banknotes than germs, however. A quick internet search might lead one to believe that just sniffing a banknote will make you high. When I taught forensic science, one year I purchased some wipes so I could perform a presumptive test for cocaine on money. A presumptive test is not definitive, but can give you a good idea as to whether or not a substance is present. (The results of a presumptive test are not admissible in court, as false positives are possible; further testing is necessary to definitively establish that a particular substance is present.) Presumptive tests are frequently performed in the field to give law enforcement a basis from which to proceed.

The particular wipes I used contained the Scott reagent, which is commonly used for the presumptive testing of cocaine. The wipes have a pinkish hue, and if cocaine were present would turn blue. The active ingredient in Scott reagent is cobalt(II) thiocyanate—$Co(SCN)_2(H_2O)_3$. It reacts readily with cocaine, which in its anhydrous form contains the salt cocaine hydrochloride ($C_{17}H_{22}ClNO_4$). The chloride ions in cocaine react with the hydrated cobalt(II) ions within the wipe to yield intense blue

$[CoCl_4]^{2-}$ ions. The test is considered presumptive because there are several ways to turn cobalt from pink to blue, with cocaine being just one of them. (Cobalt is famous for turning from pink to blue, and *vice versa*, for all sorts of reasons. Cobalt(II) chloride is well known for its use as a humidity detector—in its anhydrous form it is blue, and as humidity increases it gradually turns pink.)

I tried the wipes on a variety of banknotes, and on every one the test was negative—the wipes stayed pink. I was disappointed, hoping for a positive hit. The Scott test is very sensitive, able to detect a quantity of cocaine as low as 60 micrograms, a quantity invisible to the naked eye. Even so, my initial thought as to why all my bills tested negative was that perhaps the concentration of cocaine that was on the bills was just too minuscule to show up using this particular test.

Despite my failed attempts, research does confirm that the majority of circulating banknotes in the US do carry traces of cocaine. This contamination arises either from drug users having the drug on their hands as they handle money, or from using the banknotes themselves to snort it. In one study involving the analysis of banknotes randomly collected in cities in Massachusetts, USA, it was discovered that 67% of the circulating banknotes contained traces of cocaine.[7] Every $5 and $50 bill tested was contaminated with cocaine, but none of the $1 bills were. The amounts of cocaine detected ranged from 2 ng to 49.4 µg per bill. Evidently the method I was using to detect cocaine was just not sensitive enough; in order to detect such a minuscule concentration a gas chromatograph must be used, not a glorified wet wipe like I was using.

Other studies have shown that 95% of the bills in large cities such as Washington D.C., Boston and Detroit are contaminated with cocaine. Studies have shown much less contamination in comparably-sized cities in Asian countries such as China and Japan, while results in Canada and Brazil were similar to those of the US.[8] In the UK, virtually every banknote entering circulation will be contaminated with cocaine within just a few weeks. Contamination on euros in other European countries follow similar trends, with one study showing 100% of euros tested in Dublin, Ireland to be tainted with cocaine.[9] Another study showed that 94% of the euros in Spain contained traces of cocaine.[10]

5.13   DIRTY MONEY

Drugs and germs aren't the only contaminants on our currency. Sometimes banknotes just get plain dirty. Dirty money is more than just unsightly and unsanitary, however. It costs money. It is not uncommon for banknotes to be discarded for no other reason than being soiled. The main culprit in dirty banknotes is oil from the skin, which is attracted to the paper. Sebum, the technical term for the oils on your skin, is secreted by the sebaceous glands. These glands are connected to the hair follicles, making your hair, as well as your skin, oily.

Oxygen in the air quickly oxidizes sebum on the skin. Sebum oxidation has been linked to acne, aging and even skin cancer. It is thought that blackheads on the skin turn black owing to the oxidation of sebum, as blackheads are whiteheads before being exposed to the air. As sebum turns brown as it oxidizes, old money can acquire a dingy, brownish hue. Just as with coins, oxidation is responsible for causing money to change color.

Sebum is attracted to the surface of any banknote, as both are nonpolar. As like substances tend to dissolve in other like substances, nonpolar substances tend to dissolve in other nonpolar substances.

Cleaning old banknotes is actually a lot cheaper than printing new ones. It's not that hard to remove sebum—any number of products can remove it—but many of these products, if used on money, can wreak havoc on delicate security features, shortening a banknote's lifespan. One promising area of research for cleaning currency involves using a high-pressure form of carbon dioxide, a type of dry cleaning, if you will. Everyone knows that if you want to keep your clothes looking like brand new you will eschew the washing machine in favor of the dry cleaner. The same thinking applies to money as it is, after all, made of fabric.

Being nonpolar, carbon dioxide readily attacks nonpolar dirt and oils. However, to accomplish this task CO_2 can't be in its conventional state—it needs to change to a whole new state. By manipulating the pressure and temperature you can get CO_2 to exist in whatever form you want—liquid, gas, solid or something in-between. If CO_2 gas is allowed to escape out of a tiny opening it will turn into dry ice. If dry ice is subjected to pressures five times greater than atmospheric pressure, it will liquefy.

However, if the pressure and temperature are both simultaneously cranked up on a sample of CO_2 in a really sturdy container, a new state of matter will form, a very strange state that is not really a liquid but not really a gas either—it exists in a hybrid state known as a supercritical fluid.

Supercritical fluids tend to flow like a liquid but effuse like a gas. They expand to fill their container and can be compressed like a gas, but their density is closer to that of a liquid. Its best to think of them as a really dense gas that is under an incredible amount of pressure, or maybe as a liquid with a really low viscosity. Supercritical fluids are incredible solvents; being under extraordinary pressures they can flow right through paper and fabrics, but as they have such a low viscosity, they cause little damage.

Supercritical CO_2 is used for a variety of applications, from squeezing the caffeine out of coffee beans to cleaning greasy engine parts. Grimy looking banknotes subjected to supercritical CO_2 end up looking almost as crisp and clean as a bill that just rolled off the press. Not only are they free from dirt, sebum and drug residue, but they are also completely sterile, as even the hardiest microbe cannot survive the rigors of the supercritical treatment. It is just as effective on polymer notes as it is on the cotton-linen ones. Tests have shown that even bills stained with motor oil can be effectively cleaned using supercritical CO_2. Despite such promising results, most countries don't bother cleaning their banknotes—sorting through dirty and disheveled bills to determine which ones need a bath is a labor-intensive process. It's easier to just burn the undesirables.

5.14 WHEN MONEY WEARS OUT

According to the US Federal Reserve, the average lifespan of US banknotes is as follows:

- $1–5.8 years
- $5–5.5 years
- $10–4.5 years
- $20–7.9 years
- $50–8.5 years
- $100–15.0 years

The differences in lifespan are primarily because of how often the bill circulates. Unless you're at Disney World, you don't pass around $100 bills nearly as often as you do $10 bills. Banknotes from different countries have widely differing lifespans. In India, the average lifespan for a banknote is less than a year. The old British paper five-pound notes lasted for less than two years. Euros average about three years. In 2014, the European Central Bank, which oversees the euro, announced that it was sticking to paper for its banknotes, despite their relatively short lifespan.

The stresses of circulation will inevitably cause a weakening of the natural fibers in a banknote. Researchers for the National Bureau of Standards (NBS) in the US designed an apparatus to test the effect of crumpling on the tensile strength of banknotes. They determined that crumpling a banknote best replicated the type of wear and tear a banknote would experience in daily use. Their motor-driven automatic crumpling apparatus "alternately crumples and straightens out a banknote at a rate of seven times per minute". In a 1941 research paper, the various methods put forth to crumple paper are described:

> Various laboratory methods of wearing the paper have been tried, such as the conventional folding and bending tests, the rotating-drum tumbling apparatus, and several specially devised methods of twisting, rubbing, and bending the paper. Of the various types of treatment used, crumpling the paper seems best to simulate the type of wear which the notes undergo in circulation. It also wears the test specimen fairly uniformly over its entire surface, and has in this re-spect a distinct advantage over the other wear tests that were studied. The creasing pattern and general appearance of worn-out paper currency are fairly well simulated by the random creasing which occurs in the crumpling operation.[11]

The researchers go on to describe just how exactly their automatic crumpling machine works:

> In the crumpling apparatus the specimen is secured by means of cords passed through eyelets placed in two op-posite corners of the specimen. The square of paper is

rolled up by the fork, after which it is drawn into the crumpling cylinder, and crumpled to a small wad between the gate, and the piston. In subsequent operations the wad is expelled and opened out by the tension lever, and the smoothing arms, again rolled up and crumpled in the opposite direction, and so on.

Table 5.1 shows the data comparing the number of crumples to the tensile strength.

Today, the NBS has a more high-tech version of a crumpling tester, which basically accomplishes the same thing as the old one, only more efficiently. A banknote is rolled up into a cylinder which is then plunged with a piston that moves up and down, crumpling the bill. It works great, but lacks the charm of the original.

Once a banknote is crumpled, it takes less force to uncrumple it. The NBS devised a method to measure this uncrumpling force. They noted a measurable decrease in the uncrumpling force each time a bill was crumpled. A note crumpled twice required 4.3 kg of force to uncrumple it, while a note crumpled 40 times only required a 1.3 kg uncrumpling force.[11]

Another way to determine the quality of a banknote is to put it through the limpness test. When held by its end so it acts as a cantilever, a crisp new banknote will extend almost straight out, with minimal sagging. But every time a banknote is crumpled it will sag more and more. An aging bill will sag like a wet noodle. The NBS devised an apparatus to test for limpness. A note is harnessed into a special device and the angle with which it bends under its own weight is measured. A crisp new banknote will only bend at a 3° angle. The more times it is crumpled, the greater the angle at which it bends. Surprisingly, a bill that is

Table 5.1 Comparison of the number of crumples to the tensile strength.

Number of times crumpled	Tensile strength ($kg\,cm^{-2}$)
0	5.2
8	3.9
16	3.5
24	3.3
32	3.1
40	2.9

crumpled only once will actually bend 1° less than a brand-new bill. After that, though, additional crumples result in more bending. Eight crumples results in a 9° bend, while 36 crumples causes a bill to bend 48°.

Perhaps the most useful method for determining the quality of a banknote is to measure its air permeability. A bill is clamped into an air permeameter, which draws air through a banknote using an air pump. The more air that passes through, the greater the air permeability—a high air permeability indicates a worn bill. Every time a piece of paper is folded, hydrogen bonds within its structure are broken. As more bonds are broken the paper becomes weakened. Increased crumpling breaks more bonds, weakening the structure and causing more air to pass through. Measuring a banknote's air permeability is perhaps the best way to evaluate its condition. The good folks from the NBS could not stress enough the importance of this test:

> The hundreds of small creases that result from the crumpling of paper leave the specimen more open and more susceptible to the passage of air through the structure. Indirectly, this condition becomes a measure of the change in important characteristics of the paper. Loss in strength, increasing limpness, ability to absorb oil and grease, fuzziness of surface, susceptibility to catching dirt, and loss in the clarity of images printed on the paper, are all closely related to the loosening of the structure that is reflected in the increasing air permeability. An important advantage of this test is that it is nondestructive. Little or no change in the structure results from making the test. It is possible, therefore, for one to make repeated tests on the same specimen and to follow its behavior through a series of crumplings, noting the changes at all stages from new to completely unserviceable paper. The test is also very sensitive to small changes in structure.[11]

When worn and damaged banknotes reach a central bank, they may be slated for removal—banks make the final determination as to whether or not a banknote is fit for circulation. Most central banks utilize high-tech banknote processing systems to determine whether or not a bill is acceptable. Most likely, they use one manufactured by Giesecke + Devrient, the global leader in securities technology. One of their top-of-the-line high speed sorters—the BPS M5—uses advanced sensor technology to detect worn or damaged bills, scanning up to 33 banknotes per second. These processing systems look for any of a number of things, such as tearing, wrinkling, dirt, stains, stray marks, and so forth. In the US, if a bill has a hole with a diameter greater than 19 mm^2, it will be rejected. These scanners also test banknotes for authenticity. Older money is also removed.

A US Patent application filed by representatives from Giesecke+Devrient provides some details of the sensor technology their systems utilize:

> In bank note processing machines, bank notes can be evaluated with respect to a great variety of criteria, in order to ascertain certain properties of the bank notes to be processed... In particular, there are also employed sensors here that generate image data. Such sensors can be formed for example by line-scan cameras, which generate image data in various spectral regions that can range from the infrared *via* the visible region up into the ultraviolet region, while the respective bank note is moved past the sensor by a transport device. In addition, it is known to employ other sensors, such as mechanical sensors or ultrasonic sensors for checking bank notes in bank note processing machines. The data of the sensors are processed in a connected evaluation unit. In so doing, the mentioned properties of the bank notes, such as type (currency, denomination), authenticity, quality, state *etc.*, are ascertained from the data of one or several sensors by algorithms, and the bank notes can be sorted for example in accordance with the ascertained properties.[12]

Rejected banknotes are marked for destruction. More often than not, rejected bills are sent to the shredder. From there the shredded currency is either sent to a landfill or recycled. Little bags of this shredded money will sometimes find their way to museum gift shops or other retail outlets—the Money Museum in Chicago gives each visitor a little bag of it as a souvenir.

Shredded money has been used in a diverse array of products, in everything from mulch to insulation. When used as an additive in roofing shingles, the long cellulose fibers add strength in much the same way as straw does when used in bricks. Occasionally shredded money will even be used to make paper again; at one point the Crane Company came out with their own line of green-colored stationery made from recycled paper, dubbed "Old Money". Shredded currency has even been used as bedding for horses, a practice that had to be abandoned, however, when horses began experiencing allergic skin reactions, most likely from the ink. Some shredded money is still burned, but only in a green sort of way, as fuel for trash-burning power plants.

In the UK, almost a billion banknotes are destroyed yearly. They used to be burned, but are now recycled, often as compost. Even polymer banknotes are recycled, being turned into little plastic pellets that can be reborn as something else entirely. However, unlike old coins, which are melted down and reformed into new coins, recycled banknotes will not find new life as money again.

Recycling always shortens the fibers in paper, making them weaker. If shredded money were used to make new money, the resulting product would be inferior—putting a whole new spin on the "devaluation" of currency. Metals don't have this problem—when they are melted down any impurities are removed—the resulting product is every bit as good as it was during the first go-around.

Another problem with recycling printed material involves the ink. Before paper can be used again the ink must be removed, which can be an expensive, time-consuming process, and depending on the type of ink, not always completely effective. Currency ink is especially tough to remove.

5.15   PRIMITIVE "INKS"

We will now turn our attention to the ink—without which paper money wouldn't exist. When considering the best type of material to use for banknotes, it is important to consider how it responds to ink. If you have ever tried writing on wax paper, or any other equally nonabsorbent type of writing surface, it is an exercise in frustration. It should come as no surprise that the first people to make paper were also the first to make ink—blank paper is of little use by itself.

Ink actually predates paper if you consider primitive cave drawings. Many of these ancient markings were made using readily available organic materials such as blood or plant products. Analysis of other drawings reveal a rich mineral composition, requiring a certain degree of technical skill to formulate. After isolating a mineral, it would need to be ground into a powder and then suspended within a suitable liquid. Yellows were often made from ochre, a type of clay that derives its yellowish hue from various iron compounds, such as iron(III) oxide. Reds were often made from hematite, which contains oxidized forms of iron(III) oxide, better known as rust. Black pigments often utilized manganese dioxide. These earliest drawings were not all finger paintings either, as brush strokes reveal the use of instruments.

5.16   SEPIA INK

One of the most effective natural pigments, a favorite of the ancient Romans, is sepia, originally obtained from the ink sac of the cuttlefish (*Sepia officinalis*), a relative of the squid. Sepia is unique in that it is one of the few pigments obtained from an animal. Like the squid, when threatened cuttlefish will squirt out a cloud of ink, making the water opaque, allowing for a quick getaway. Sepia ink owes its color to melanin, the same pigment that colors our hair, eyes and skin. Melanin makes up about 15% of the mass of cuttlefish ink, with a typical ink sac containing about a gram. Sepia ink also contains a fair amount of amino acids as well as salts containing ions of iron, calcium, magnesium, potassium and sodium, along with heavier metals such as cadmium, copper and lead.

Sepia ink has endured for two millennia. Recipes to make it abound. One 19th century method is as follows:

When the creatures are captured, their glands are carefully extracted and sun-dried so as to solidify the contents. In this state ink bags are sent into commerce. The colourman subjects the sacs to boiling in a solution of soda or potash, whereby the colour is dissolved out of the receptacle, and being filtered clear of all fragments of the animal tissue, is next precipitated by the addition of acid, collected on a filter, washed, and dried. It then forms an exceedingly useful pigment.[13]

Natural sepia ink was a common site in artist studios well into the 19th century, but has largely been replaced by synthetic pigments. Natural sepia ink can still be purchased, though. The dark brown pigment in sepia ink is incredibly resilient, making it one of the most durable natural stains. It can make water completely opaque in concentrations as low as one part per thousand.

The longevity of sepia made it a natural choice for use in early photography—most photos you see from the latter half of the 19th century are sepia toned. Early photos tended to fade quickly owing to rapid degradation of the metallic silver, which was used to create the photographic images. As silver is oxidized rather quickly by atmospheric gases into colorless silver ions, old photos faded and sometimes completely disappeared, but silver sulfide is remarkably stable. As sepia ink just happens to contain sulfide compounds, the addition of some of this sepia ink during the developing process would convert metallic silver into silver sulfide, making these sepia photos timeless. Today sepia-toned images are primarily created for effect.

Since their beginning, banknotes have contained a signature of a bank or treasury official, verifying their authenticity. These signatures were not printed, however, but were signed by hand in ink. Even though no longer handwritten, this tradition has continued to the present day, with almost all banknotes containing a signature. On older banknotes, you will often see signatures written with brown ink, adding another layer of security. You will sometimes see older banknotes advertised as containing "sepia" signatures, which usually refers to the hue as opposed to a specific chemical composition. Other types of ink on old documents can also be brown, most notably iron gall ink.

There are a variety of methods that can be used to determine the chemical identity of an ink signature. Melanin-based inks are difficult to identify—the very characteristics that make them useful as an ink make them difficult to analyze using traditional wet-chemistry methods, as their components are notoriously insoluble, resisting degradation. The heterogeneous nature of these inks also makes them difficult to purify. The positive identification of sepia can be accomplished with more sophisticated analytical techniques, but many of these are destructive in nature. Fortunately, there are now a number of non-destructive techniques for ink identification that can be performed quickly and easily, which will be discussed later.

5.17   CARBON-BASED INKS

Even though brown sepia ink has its place, through the ages most types of ink have been black. Ink as we know it today traces its origins to both China and Egypt, which appear to have simultaneously developed ink around 2500 BC. Every form of ink is composed of two main components, the coloring agent and a liquid vehicle to suspend or dissolve it. A pigment differs from a dye in that pigments are insoluble in the vehicle, forming a suspension. Dyes, on the other hand, are soluble in the vehicle, forming solutions. Early forms of ink relied almost exclusively on insoluble pigments.

The earliest forms of ink typically utilized lampblack, a finely divided form of carbon, to make what is commonly called India ink, although it was actually invented in China. India ink is still used today by art students who want to produce striking ink drawings made from the blackest black. The formula for India ink has undergone numerous iterations in its history, but they all use essentially the same pigment, the difference being the type of binder used. Shellac and borax are often added to help keep the pigment in suspension—shellac and borax react to form soap, an emulsifier that will keep the carbon particles from settling out.

Lampblack was originally obtained by scraping away the soot from the inside of oil lamps. Lampblack has a variety of uses, being found in copier toner, tires, plastics and just about anyplace else in which a rich black color is desired.

Lampblack makes an excellent pigment for ink, not only because of its opacity but also because of its ability to withstand degradation from light, oxygen or other chemical processes. As soot is a product of combustion you can count on it being stable; water and carbon dioxide, the other products of combustion, are stable as well. The products of a combustion reaction are always in a much lower energy state than are the reactants. Charcoal is notoriously difficult to light for that very reason, as charcoal is essentially burnt wood that has already been through the fire and is not keen on going through another one. As it has already been oxidized additional oxidation is difficult, making charcoal hard to burn—it's like trying to get rust to rust some more.

Therefore, any ink made from an oxidized form of carbon is incredibly stable—ancient Egyptian papyrus manuscripts dating back as far as 2500 BC are still readable today owing to the stability of the carbon-based ink used to write them.

Soot was not the only source of black pigment in ink. Bone black was also commonly used, which was made by burning animal bones in limited oxygen. Rembrandt was a big fan of bone black, being convinced it produced the deepest black of any pigment.

Although black may have been the preferred ink color throughout history, on special occasions red ink would be used, which was commonly prepared using mercury(II) sulfide. Mercury(II) sulfide is obtained from the mineral cinnabar, and is famous for making vivid vermilion hues. With a K_{sp} of 4×10^{-53}, its extreme insolubility makes it perfectly suited for pigmentation. Over time, oxidation will convert it into a darker form. Red ink was often used for emphasis or to denote important dates and events, hence the term "red letter day". Traces of mercury(II) sulfide have even been found in the red ink of the Dead Sea Scrolls, which were written between the 4th century BC and the 1st century AD. The majority of the ink used to write the scrolls, however, was made from lampblack derived from burning olive oil.

As further evidence of the stability of carbon-based inks, one only has to consider the recent find of a one-kwan banknote issued by the Ming Dynasty, dating back to around the year 1375 (Figure 5.4). It is one of the world's oldest surviving banknotes, being discovered folded up inside the head of a wooden Buddhist sculpture. It was printed using a carved woodblock on thick handmade mulberry bark paper with black ink, but also contained two vermilion seals. It was much larger than today's banknotes, having dimensions of 33.5×22.5 cm, about the size of a sheet of copy paper. These banknotes were printed from ink containing lampblack, obtained by burning the oil from fir trees.

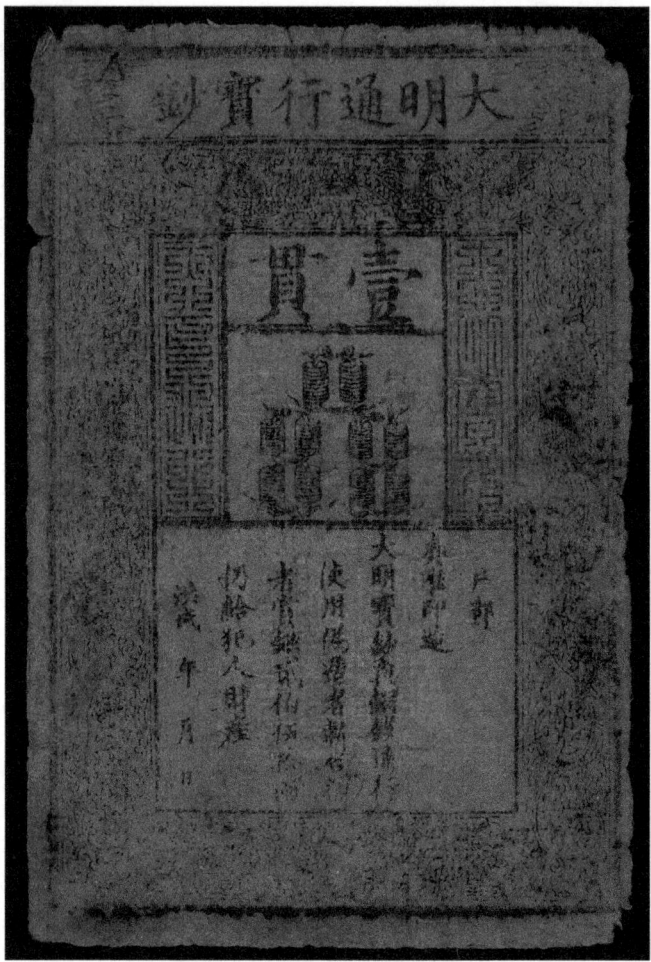

Figure 5.4   Bank Note of the Ming Dynasty.
© The Trustees of the British Museum.

Whenever viewing old manuscripts in a museum, I am always struck by the blackness of the ink, which enables 2000-year-old manuscripts to still be eminently readable.

Lampblack makes an excellent pigment in large part because of its tremendous surface area. Owing to its small particle size, it forms a colloidal suspension when added to water. As a colloid is not a true solution the lampblack needs some help staying suspended. Gum arabic—the dried sap from the acacia tree—has long been used in carbon-based inks. It not only holds the suspension together, but also increases its viscosity.

Gum arabic is a natural glue—not only does it stabilize the ink, it binds it to the paper. It is more than a little important that the ink stick to the paper, which we tend to take for granted when we write with an ink pen. If you dissolve a little gum arabic in water and touch it, you can clearly see for yourself just how sticky it is. The next time you eat a sticky type of candy, like gum drops or gummy bears, look at the ingredients—there's a good chance you will find gum arabic listed among them. We tend not to think of ink as sticky because we avoid touching it, but if you have ever accidentally gotten your hands in a batch of ink you will quickly find out just how sticky it is—if you ever run out of glue, ink can serve as a worthy substitute.

Colloids are sticky because of their huge surface area. An increased surface area leads to greater intermolecular attractions between objects in contact. These weak bonds are generally known as van der Waals forces, or if they are between nonpolar substances London dispersion forces. The stickiness of everything from gecko's feet to Elmer's glue is a result of these intermolecular forces of attraction between neighboring atoms. As electrons swirl about the nucleus, it is highly unlikely that the electron cloud of any one atom will be totally symmetrical at all times. The tiniest imbalance will make one side of the atom a little more negative than the other side, setting up a temporary dipole. This temporary dipole will then in turn induce a dipole in a neighboring atom, and so on, until a very predictable result ensues, namely stickiness caused by the electrostatic attraction between unlike charges—a direct result of totally random and utterly unpredictable behavior.

The lickable adhesive on the back of an envelope flap is most likely made from gum arabic, and for many years served as the glue on the back of postage stamps as well, eventually being replaced with polyvinyl alcohol, which has since mostly given way to pressure sensitive adhesives. If you have ever tried to remove an old-fashioned stamp from an envelope by peeling it off, it's not easy—you will either seriously compromise the stamp or the envelope if you attempt to remove it. However, if you soak the envelope and attached stamp in water it will come off cleanly, as gum arabic is soluble in water.

One big problem with carbon-based inks is that because they don't penetrate very far into the paper they can be rubbed off with minimal effort. This drawback was not necessarily viewed as a terrible thing in the earlier days of writing, however, as the easy removability of the ink meant that one could reuse the parchment (animal skin) or paper, which was much harder to come by than the ink.

5.18   IRON GALL INK

Eventually carbon-based inks gave way to, of all things, ink made from the galls of oak trees. If you look at any oak tree, it is not hard to find galls on the leaves and stems. These spherical growths represent the plant's response to a wasp that lays eggs on developing leaf and stem buds. When the eggs hatch, the larva secrete growth regulators that stimulate the oak tree to form a protective casing around it. Within each gall is a developing larva—if the gall has a hole in it the larva has already turned into an adult wasp and departed. The galls contain tannic acid, and as the concentration is always highest before the larva has matured, it is best to try to harvest the galls while the larva is still present.

Tannic acid derives its name from the Latin word *tannum*, meaning oak bark. Oak galls are a rich source of tannic acid, being traditionally used to tan leather. The oak galls contain a specific type of tannic acid known as gallotannic acid. Oak galls have been highly sought after as a treatment for everything from hemorrhoids to ingrown toenails. The galls are also effective for coloring textiles if a deep black color is needed.

The development of iron gall inks paralleled advances in papermaking, primarily that of flax and linen papers. Iron gall ink was hugely popular for about 1500 years, even into the 20th century. Making iron gall ink is fairly simple, requiring only iron(II) sulfate, some ground up oak galls, and a suitable binder, such as gum arabic. It is still used today, though its use has mostly been supplanted by more modern synthetic inks.

An 1852 recipe for making iron gall ink lists all the necessary steps:

A good writing ink is made as follows: Take six ounces of finely bruised galls, four ounces of gum Arabic, four ounces of green vitriol, and six pints of soft water. Boil the galls in the water; then add the other ingredients; and mixing the whole well together, keep it in a well-corked bottle, occasionally shaking it. In two months' time carefully pour off the ink from the residue into glass bottles, which should be well corked; a few drops of cloves or creosote, put into each bottle, prevent moulding. The addition of a little sugar to ink gives it a gloss.[14]

The iron(II) sulfate reacts with gallotannic acid in the oak galls to form the iron(II) tannate complex, as well as a little sulfuric acid. Iron(II) tannate is water soluble and can easily work its way down below the paper's surface. As it is colorless in solution, iron gall ink is the original invisible ink. In order to see what you wrote, you would need to wait until some of the iron(II) tannate oxidized into iron(III) tannate, which would be a deep bluish-black color. If you waited until you could see the ink to write with it, however, you would have problems. As iron(III) tannate is not water soluble it wouldn't penetrate into the paper as deeply as iron(II) tannate. The best ink would be a mixture of old and new ink—that way you could see what you were writing but could also achieve good penetration. Over time the purplish iron gall ink fades into a brown color.

If a manuscript or drawing was inked between the 13th and the 19th century, there is a good chance it was made using iron gall ink. Unlike the grittier carbon-based inks, it was prized for its smoothness and ease of application. The drawings of Rembrandt and Leonardo da Vinci were made from iron gall ink, as were the compositions of Bach. Thomas Jefferson even penned the Declaration of Independence with it. However, owing to its acidic nature, iron gall ink can have a corrosive effect on paper, as its pH is between 1–3 (Figure 5.5).

Many historical documents have been literally eaten away by iron gall ink. Documents fare better if the ink is applied with a brush as opposed to a pen, as a pen drives the ink more deeply into the paper, where it can do more damage. As there were hundreds of recipes for iron gall ink, some producing a more acidic ink than others, not all manuscripts penned with this ink are in mortal danger. However, it's not only the acid that degrades the paper—the iron sulfate compounds are corrosive in

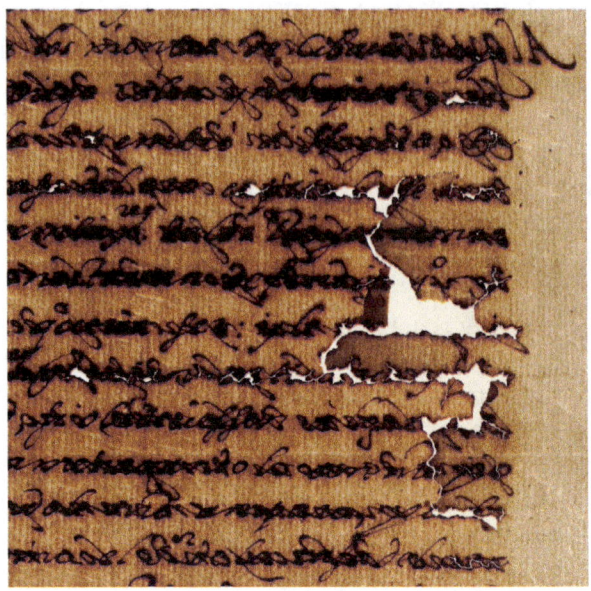

Figure 5.5 Corrosive effect of iron gall ink on a 16th century manuscript.
Credit: Traveling Scriptorium from Beinecke MS 128, Beinecke
Rare Book & Manuscript Library.

their own right, with vitriol being the Latin name for a sulfate
compound. One promising conservation method involves treat-
ing documents containing iron gall ink with calcium phytate, a
complexing agent which removes the iron ions. This treatment is
followed by a wash with calcium bicarbonate, which not only
neutralizes the acidity but also helps to reform hydrogen bonds
in the paper, restoring its strength.

During the US Civil War of the 1860s, the Confederate
States of America (CSA) printed their own currency. As they
were cash strapped, most of their money was printed on rice
paper, a thin and almost waxy type of cheap paper. Owing to its
texture it was difficult for ink to penetrate it. As these bills
needed to be signed, iron gall ink was chosen owing to its excellent penetrating ability. As the paper was so thin, one way to
determine if Confederate currency is genuine is to look for
bleeding of the ink through the paper. Like sepia, iron gall ink
was not used as a printing ink but rather as a writing ink. Its use
in currency will therefore be limited to written signatures on
the banknotes.

5.19 ANALYZING INKS

The best nondestructive way to determine the type of ink on an ancient document is to use multispectral imaging, which involves exposing an ink sample to various types of light outside the visible portion of the spectrum and then analyzing the response. A special camera records the image, allowing you to see what wavelengths are absorbed. Filters can be employed to block out certain frequencies so as to ascertain what particular wavelength a pigment might be responding to.

Alternative light sources are great at detecting document forgery. I once had a student bring to me a paper I had recently returned to him. It had a grade written across the top, different than the one I had originally assigned. The original grade, which was written with red ink, was blotted out with a heavy marking of blue ink. A new grade was then written down, in handwriting that did not match my own. On a whim, I placed the paper under a blacklight, which clearly revealed the old grade underneath the inked-out portion. The ink the student used to cross off the grade was transparent to ultraviolet light, but the ink I used to write the original grade was not, being clearly visible. The blacklight allowed me to see through one type of ink to reveal another type underneath.

The same principal is essentially at work when we receive an X-ray. Soft tissue is transparent to X-rays but bones are not. In the same way, different types of ink will respond differently to different types of light. Carbon-based inks, for example, are strong absorbers of infrared radiation. When a multispectral image sensor records an image of this ink, the resulting image it produces will allow you to easily spot it, as it may appear much "blacker" than other types of ink. Iron gall ink will absorb infrared radiation differently than carbon-based inks.

For a more precise determination, infrared spectroscopy can be used, which can reveal the characteristic infrared (IR) signature of each type of ink, as measured using differences in their infrared spectra. The underlying premise is that the various components in different types of inks absorb specific frequencies of IR radiation at differing levels, as well as responding differently to different types of IR—one pigment might be an excellent absorber of near-IR but a poor absorber of far-IR.

Our eyes perform an analysis of each substance's ability to absorb visible light when it determines its color.

A related technique is known as Raman spectroscopy, which analyzes the vibration of molecules in response to irradiation by a powerful laser. By measuring the pattern of scattered light that emerges as a result of these molecular vibrations, a characteristic spectrum of a substance can be constructed. Every substance will vibrate a little bit differently, and thus scatter light a little differently.

Analyzing the responses of pigments to varying wavelengths of electromagnetic radiation, by any number of methods, is one of the most commonly employed techniques used today to detect forgeries in manuscripts, artworks and banknotes. The paper as well as the ink can be analyzed, and as substances change over time, the age of a document can be inferred as well. In 2005, in Leonardo da Vinci's "The Virgin of the Rocks," the presence of several additional drawings was discovered underneath the painting, being revealed through infrared imaging. Hundreds of such examples of hidden drawings and paintings have also come to light since the advent of this and related technologies that allow us to peer through one layer and see the one underneath.

5.20   RESINS

With advances in printing technology in the 15th century and with it the increased dissemination of printed materials, the union of ink to paper became more important than ever. One of the biggest innovations to ink was the addition of oil, at first linseed oil, to serve as the vehicle. As flax was already a key ingredient in papermaking, as the raw material for linen rags, it was only fitting that the oil derived from its seeds be used for the ink. Adding linseed oil greatly increased ink's viscosity, enabling it to adhere to the metal printing plates. Ordinary ink used for writing would have been much too thin and watery for printing.

Although the pigment and the vehicle have undergone changes over the years, by far the biggest changes in printing inks have involved the crucial third component, which is referred to in the printing industry as the resin. Resins have a host of duties, chief of which is to hold together the pigment and the vehicle. They typically act as an emulsifier, keeping the pigment particles in colloidal suspension. Resins also serve to increase viscosity, aiding in flow and dispersal, as well as helping bind the ink to the substrate; they also regulate drying time and determine the final look and feel of the ink after its application. A high molecular mass resin will evaporate more slowly than one of lower molecular mass.

The first resins used in inks were derived from actual resin—the highly viscous yellowish sap-like substance that can sometimes be seen extruding from the trunks of trees. As resin is highly nonpolar, it is only soluble in a nonpolar solvent. One of the key ingredients in resins is terpenes—volatile hydrocarbons made up of various combinations of isoprene (C_5H_8). Terpenes are why pine trees smell piney and lemons smell lemony. The distinctive smell of marijuana is caused by terpenes; so are the odors of menthol and eucalyptus. If you remove the terpenes from resin you get turpentine. The leftover solid is called rosin, which you might apply to the bow when a violin is played.

Today the term resin is used to refer to any binding or dispersal agent used in ink. Resins in ink today typically involve a mixture of synthetic polymers. As resins have evolved, so have the solvents—although volatile hydrocarbons are still employed, ketones, ethers and a host of other liquids may also be used. If cake batter was ink, think of the powered cake mix as the pigment, eggs as the resin and milk as the solvent.

5.21   THE AROMA OF INK

Volatile solvents and resins combine to give ink its unique smell. Every type of ink will have its own distinct aroma, owing to the evaporation of these volatile compounds.

As there is wide variation in the volatility of compounds, they can sometimes be used to determine the age of ink on printed material. Unless you are left handed, you usually don't experience smudging when using a ballpoint pen as the ink dries so quickly. However, ink doesn't dry this fast without a little help—ballpoint pen ink often contains 2-phenoxyethanol, an aromatic alcohol that hastens the drying process. The low-cost clear Bic ballpoint pens contain this substance, and it is partly responsible for the ink's pleasant odor. This volatile compound has been used as a way to determine the age of ink, and even though the ink seems dry, evaporation of this alcohol is still occurring; the evaporation rate is especially rapid over the first 6 to 8 months of its application, and then slows down after that, with little evaporation occurring after 2 years. Using a gas chromatograph/mass spectrometer, it is possible to determine the amount of 2-phenoxyethanol in an ink sample. It is therefore possible to compare two ink samples, and determine with reasonable certainty whether or not two writing samples were written at the same time. You can also determine the age of a particular ink that has been applied to paper.

An added benefit of 2-phenoxyethanol is its ability to attract termites, as it mimics a common pheromone that termites use to communicate. If you want to know if your pen ink contains 2-phenoxyethanol, make a mark on a piece of paper and see if some termites will toe this line. If termites follow the trail of your pen ink, it most likely contains 2-phenoxyethanol.

The odors of both paper and ink are cause by the emission of volatile compounds. Unlike ink, the older the paper the more likely it is to smell. The smell of an old book is mostly due to the paper—you are probably getting a good whiff of vanillin, owing to the breakdown of lignin. Cellulose within the pages also degrades, releasing acetic acid and furfural (which is why coffee smells so good). The smell of a new book, on the other hand, is not due to the paper but is primarily caused by the ink (and the adhesives). If you walk into a new bookstore you will be greeted with an entirely different set of odors than those that await you in a used bookstore. Just like a new car smell dissipates over time, so too does the smell of new books.

Unlike new coins, which have no odor, the volatile compounds in the ink impart a distinct odor to new paper money. Analyses of these odors have revealed the building blocks of currency inks, with aliphatic aldehydes and alkanes at the top of the list. Despite the distinct aromas of new currency, government agents and bank personnel are not trained to use odor as a way to differentiate real bills from counterfeit ones, as it doesn't take long for a new bill to pick up the aromas of its surroundings, and begin to smell like body odor, or worse. Dogs, however, can be trained to smell money.

In their efforts to stop the flow of cash from drug smugglers, the US Customs Service has employed currency-sniffing canines to locate large caches of cash. Even though a dog's nose contains only 50 times as many olfactory receptors as ours, their sense of smell is 100 000 times more acute. In training these dogs, a minimum odor threshold is established so they won't be alerting authorities to every passerby who has a few bucks in their pocket. The dogs are trained using both circulated and uncirculated banknotes, as well as shredded currency and even the ink that US banknotes are printed on.

An analysis of currency ink undertaken by the US Customs Service Research Laboratory revealed that of the six inks used in US currency the ink that smelled the most like money was intaglio green (I286). Intaglio black magnetic ink had a similar odor to I286, but was stronger. Intaglio black nonmagnetic ink (I600) had a slight aldehyde odor, while currency overprinting equipment (COPE) ink (I347) smelled like licorice. COPE green (I347) had a pungent acidic tang, and another COPE green (I1201) smelled sweet.[15]

The dogs were also able to detect Canadian currency, even though they were only trained with US currency. Even though the ink formulations are different for Canadian currency there is evidently enough of a similarity for the dogs to pick up its scent.

The Customs Service found that the solvents used in currency ink evaporated fairly quickly, which is to be expected, as they are rather volatile. Researchers also found that the ink curing compounds lasted the longest, and it was these compounds that were the ones most likely to be detected by the canines. The ink used in US banknotes never fully dries, and even bills three decades old still show significant amounts of these curing agents; as these substances are among the longest lasting in currency ink, they are most likely to produce the odors the dogs detect.

5.22 WATCHING INK DRY

Much like newspapers, the ink used to print banknotes is absorbed by the substrate, and as most of the ink does not remain on the surface its lack of complete drying does not pose an issue. You can easily demonstrate this property yourself by taking a US banknote and rubbing it firmly across a sheet of white paper a few times. You will see a smudge of black ink if the face of the bill is rubbed against the paper and a green smudge if the back is rubbed. Newer bills leave a bigger smudge than older bills. The euro, however, doesn't leave a smudge at all owing to the protective varnish on its surface.

Not-completely-dry ink is not all that uncommon; take any piece of printed matter—whether it be a book, magazine or photocopy—and rub it across a piece of white paper. The ink will smear. Ink that dries too quickly poses any number of problems. If ink dries too rapidly it will be nearly impossible to apply—unless it's on paper, dry ink is worthless. Ink that dries too fast can clog up the ink delivery system, be it a printing press or a ballpoint pen.

Some types of inks, however, do need to be fast drying and do need to dry completely. A driver's license should not smear when you rub it. As there is essentially no ink penetration into the substrate on plastic or other types of glossy surfaces, the ink must dry completely or it would simply run off. The new banknotes made of polymers utilize fast-drying ink. If a polymer banknote is rubbed against a sheet of white paper you will not see ink smudges.

Ink dries in one of two ways: either through oxidation or evaporation, usually, it's a combination of both. When the first printed materials rolled off the printing press the linseed-based ink first penetrated into the paper, and whatever was left behind was oxidized. As oxygen reacted with the oils, polymerization occurred as long-chain fatty acids formed. The cross-linking of the polymer chains resulted in a hard, durable film that trapped the pigment particles, binding them to the substrate. As cross-linking continues to occur over time, ink can become brittle, leading to cracking and peeling. Not all inks rely on oxidation to dry. Inks that are water based typically dry by evaporation, just like water-based paints.

5.23   INK PENETRATION

For the inks that don't completely dry, the ink must penetrate deeply into the paper. Penetration occurs as a result of capillary action. You can observe capillary action by sticking a piece of celery into blue colored water—in no time the piece of celery turns blue as water climbs up the walls of the tiny xylem tubes.

Capillary action works as a result of the combined action of adhesive forces, which enable water to stick to the sides of the tube, and cohesive forces that enable water molecules to stick to one another. There are all sorts of little channels between the cellulose matrices in paper—fluid is drawn into these channels by capillary action. In a very porous paper there will be more channels, and thus better absorption. Paper towels rely on capillary action to do their job, being intentionally designed to contain a lot of tiny tubes so they can absorb a lot of water.

Sufficient penetration of ink into paper is necessary to make sure the ink stays put, and is especially important if the ink never dries completely. However, if the ink penetrates too deeply that is not good either. Try writing on a paper towel with a ballpoint pen—you will see a lot of bleeding, which you would expect, as the paper towel is doing what it was made to do—absorb liquids *via* capillary action. Some capillary action is good, but too much is undesirable—the ink should penetrate just enough to stay permanent, but not so deeply that your markings look like a chromatogram.

To control how much water penetrates the surface of paper, sizing agents are added during the papermaking process. Sizing involves adding substances to paper in order to decrease its permeability to water, as well as adding a layer of protection so the surface is less fragile. Paper that has undergone the most sizing will be the glossiest, like a milk carton or juice box. Paper towels are unsized, as is blotting paper, coffee filters and laboratory filter paper. Unsized paper will typically lack starch, as starch is a very common sizing agent. If you want to know how much your paper has been sized, try using it to soak up a spill. A piece of copy paper is basically worthless, with yellow legal pad paper or white notebook paper being a little better, as they are designed to be written on with ball point pens and the like—as they need to absorb ink, they contain less sizing.

Sizing agents can be added at various stages of the papermaking process. Paper that is wet-sized will have the sizing agents impregnated throughout. Surface sizing occurs when sizing agents are applied to the paper while it is in the process of drying, and thus only penetrate the surface. Starch and gelatin are two common surface sizing agents. If starch is used, however, it must be modified to make it amphoteric, so that the hydrophobic end is oriented to the outside of the paper and the hydrophilic end to the inside. The garden variety unmodified corn starch would not work as a sizing agent, as it tends to absorb water.

Newsprint undergoes very little sizing, which is why it tests negative for starch, as we discovered earlier. To print papers at high speed without smudging, the paper must absorb the ink quickly. With such a rapid rate of absorption, the lack of complete drying is not an issue, as it is dry enough and won't come off unless you work at it. Newsprint is not a great medium for a lot of inks—if the proper ink isn't used there is considerable bleeding.

Sizing agents are used in banknote paper, but these agents are never made from starch. Cotton or cotton-linen banknotes still absorb water, but not as much as unsized paper. It is easy to test the absorbency of a banknote by recording its mass before and after you soak it in water. I measured the mass of a US $1 bill, which was 0.98 g, and then soaked it in water for 30 minutes. After blotting off the excess I measured the mass again, which was a somewhat surprising 1.43 g. I didn't think the dollar bill would absorb quite this much water—almost 50% of its mass. However, this high rate of absorbency makes sense if you consider that the substrate must be absorbent enough to take in a considerable amount of ink, which after all doesn't dry completely.

Sizing agents are also used to prevent counterfeiters from removing the ink from a legitimate banknote and then printing a higher denomination banknote on the same paper. To this end the Central Bank of the Bahamas began issuing notes that were coated with a special sizing agent, making it considerably more difficult for counterfeiters to expunge the ink.

5.24   PRINTING MONEY

Now that we've looked at the paper and the ink, we will see how it all fits together to make money. One of the oldest types of printing processes is relief printing. If you have ever been fingerprinted, you have taken part in relief printing—the raised ridges on your fingertips get covered in ink and are then transferred to paper. If you take a wood block to a stamp pad, and then stamp an image, you have printed in relief. The very earliest Chinese money was produced using woodblock printing. The old-fashioned printing presses, in which individual raised letters were set in place and then inked, printed images using relief. A lot of American colonial money was printed using raised type, providing a cheap and efficient way to print large sums of cash.

Relief printing then gave way to intaglio printing, which has traditionally been associated with the printing of banknotes, as it can capture extremely fine details, as well as produce raised lettering. Most early European money was printed using the intaglio process. The word *intaglio* (silent g) comes from the Italian word *intagliare* for engrave. Intaglio printing is the opposite of relief printing, in that the image to be printed is below the surface of the printing plate. Ink is then applied to these recessed areas. Considerable pressure is required to force the ink onto the paper, which is furnished by a roller press—a cylindrical drum that the paper passes under.

The first step in intaglio printing is to produce the recessed plates. Markings can be engraved into a copper plate using several different methods. A steel cutting tool, known as a burin, can be used to engrave a plate. Sometimes acid is used. A thick waxy substance impervious to acid is applied over the entire plate, and then an image will be engraved into the waxy coating. When dipped in acid, the acid reacts with the exposed copper, etching an image into the plate. Plates prepared by the action of a chemical process are known as etchings, as opposed to engravings, which are inscribed manually.

Nitric acid has often been used to etch plates made of copper, as it reacts so readily with copper. However, the reaction is so violent that it is difficult to control, so a more manageable substance, known as Dutch mordant, has often been used in its place. Dutch mordant is a mixture of hydrochloric acid and potassium chlorate. Hydrochloric acid does not react with copper by itself, but will if mixed with potassium chlorate. Dutch mordant was used by Rembrandt to etch a number of copper plates, with remarkable detail. His recipe consisted of 1 liter of water, 125 mL hydrochloric acid, 25 g potassium chlorate and 25 g sodium chloride. The acid liberates the chlorine from potassium chlorate, which in turn attacks the copper. As Dutch mordant does not react as fast as nitric acid, it is the preferred method of etching if fine detail is desired. Iron(III) chloride is also a commonly-used etching material, as it too reacts with copper.

5.25   MAKING MONEY

All British banknotes are printed by De La Rue, a private firm that bills itself as supplying a third of the world's banknotes, both paper and polymer. They have a long and rich history of printing everything from playing cards to postage stamps. They not only print money for countries all over the world, but also supply the paper. The euro is printed by several different printers, including De La Rue. The world's largest producer of banknotes is the China Banknote Printing and Minting Corporation (CBPM), a state-owned company that not only prints all of China's money, but that of numerous other countries as well. All US currency is made by the Bureau of Engraving and Printing (BEP), which has locations in Washington, D.C. and Fort Worth Texas. The printing of banknotes is becoming ever more centralized, with a number of nations opting out of printing their own money. Money has become increasingly complex—the incorporation of ever more security features has made it more cost feasible for many of these countries to contract out the printing to an established entity.

When paper money first came on the scene, it was typically printed in either relief or intaglio. The printing of today's money does not rely on any single technique, but rather employs several different methods. We will examine briefly the steps used in printing American money, and although every country that prints money or has it printed will do things a bit differently, the overall process is similar. The first step is always to design the banknote. This design must then be engraved on a die, which is usually made from soft steel. (As you may recall, dies used to strike and stamp coins are always made from hardened steel, as they undergo significantly more wear and tear.) The engraved design is a mirror image of the original design, so it must be cut in reverse.

The entire design is not printed on a single die, but each part is rather printed on one of several different dies which will eventually form the entire image. A master die is then made from each of these dies. The master die is used to make a number of plastic molds that are then grouped together to make a plastic master plate—in the US it is eight molds long by four molds wide. This plastic plate is used to create a number of identical metal printing plates using some ingenious electrochemistry.

At one of the BEP money factories, the plastic master plate is coated with a solution of silver nitrate and then immersed in a 2600-liter tank of nickel salt solution through which an electric current is applied. After about 17 hours enough nickel ions will have deposited onto the substrate to produce a layer 1 mm thick. The plate is next dipped in a hexavalent chromium bath for 45 minutes to add an 8-μm chromium layer for protection. The metal plate is then separated from the plastic mold. Each plate is used about a million times before it is retired.[16]

These metal plates never come into contact with paper, however, nor are they ever stained by ink. The image on the metal plates is transferred to a large rubber roller in a process known as offset printing, as the image on the plate is "offset" onto a rubber roller, hence the term "rolling off the press". Offset printing is also known as lithography, and thus every American banknote is, at least partly, a lithograph.

Lithographs originated with flat plates of limestone—*lithos* is the Greek word for stone. An image would be drawn on the plate using wax or another nonpolar substance. Acid would then be used to etch around the image; limestone was used because it reacts readily with acids. The acid would not only create a recessed area surrounding the waxed image, but would also leave behind a water-soluble salt, which would help to attract water. If nitric acid was used for etching, calcium nitrate would be left behind. The plate was then coated with water containing gum arabic, which stuck to the salty limestone but was repelled by the wax. When an oil-based ink was applied to the print, it would be repelled by the etched area containing water, but would be attracted to the nonpolar wax.

To offset the image onto a roller, first an image of the plate must be burned onto the roller. Using a laser image sensor, a negative image of the printing plate is made on light sensitive film. This negative is then placed on a thin sheet of steel that is coated with a light sensitive polymer. When exposed to ultraviolet light, some areas of the film allow light to pass through while other areas do not. If light passes through the negative it exposes the underlying light-sensitive polymer coating, thus creating an image. The polymer coating on the unexposed areas is then removed.

When ink is applied to the plates, it will adhere to certain areas but not to others. The polymer coating is oleophilic (oil loving), allowing the ink to adhere only to this portion of the plate.

The cotton-linen fabric on which the banknotes are printed will already have had the security features such as watermarks and security threads embedded within the paper when the ink is applied. Ink application requires a pressure of about 138 000 kPa, which is about 1360 times what the atmosphere exerts. The backs are printed first at a rate of 10 000 per hour; after drying for 72 hours the faces are printed using the same process.

The offset process is only used to print the background colors. The intaglio process is used to print the portraits, border designs and the denomination. This process involves creating an engraved plate and then applying ink only to the engraved, or recessed areas. Although offset printing is fine for most everyday printing jobs, it cannot capture detail like intaglio printing can. All microprinting on currency is done through the intaglio process.

Although the face is mostly printed with black ink, all US currency is printed with green ink on the back, thus the slang term "greenbacks" for American money. This green ink has been in use since 1861, and was originally used to prevent counterfeiters from taking pictures of money and using these images to make fakes. As most 19th century cameras only utilized black and white film, you couldn't replicate green money by taking its picture. Even though counterfeiters eventually found other ways to duplicate the green ink, the tradition stuck. For quite some time, it was illegal to film or take a picture of money in the US. If you look carefully at portrayals of money in old movies, you can see that it's not genuine—film studios often used pesos in place of dollars, or used prop money instead of the real thing.

The original green ink for American dollars was made from chromium(VI) oxide and linseed oil, coincidentally the same type of chromium used to electroplate the nickel plates today. The chromium(VI) oxide was mixed with linseed oil, making a very durable, insoluble ink. The black ink was carbon-based, making the whole recipe not that different from the ink used to print some of the very first banknotes. The exact formulation of the inks used by the BEP today, however, is a closely guarded secret.

All banknotes up to this point will be identical until they are printed with a serial number. Every US banknote has a different serial number, and these are printed using the letterpress technique, in which a raised surface is inked and then printed on the bill. The Treasury and Federal Reserve seals are also printed using the letterpress technique, as are the Federal Reserve identification numbers.

The next step is to cut the sheets into individual bills—there are typically 32 bills per sheet—which are then banded together in groups of 100 banknotes each. From there they are bundled and shrink wrapped in groups of ten. Four bundles are grouped together to make a brick, which is shrink wrapped again. Each brick contains 4000 banknotes. As the shrink wrap is heated, it fits tightly over the bills. During heating the polymer strands making up the plastic shrink wrap return to their original coiled orientation, creating shorter strands.

Each year billions of new banknotes are printed worldwide—in 2017 11 billion euros alone were printed. The cost to make these banknotes is quite cheap—in the US it costs about a nickel to make a dollar, a great return on your investment. The innovations that have gone into banknotes are astounding—they come in a dizzying array of sizes, shapes, colors, and increasingly, materials. Some banknotes have holes, just like you might see in some coins. Most banknotes have intricate security features woven into each bill. Even though banknotes have emerged as an art form in their own right, the driving force behind almost every innovation is an attempt to deter counterfeiting, which we will examine in more detail in the next chapter.

REFERENCES

1. C. K. F. Hermann, The shrinking dollar bill, *J. Chem. Educ.*, 1997, **74**(11), 1357.
2. P. Hwang, How often does a banknote change hands? On Stride Financial, 2014, https://www.onstride.co.uk/blog/often-banknote-change-hands/.

3. F. Vriesekoop, C. Russel, B. Alvarez-Mayorga, K. Aidoo, Q. Yuan, A. Scannell, R. R. Beumer and G. Menz, Dirty money: An investigation into the hygiene status of some of the world's currencies as obtained from food outlets, *Foodborne Pathog. Dis.*, 2010, 7(12), 1497–1502.

4. H. Gedik, T. A. Voss and A. Voss, Money and transmission of bacteria. *Antimicrob. Resist. Infect. Contr.*, 2, 2013, 22.

5. P. T. W. Capt, P. T. Ender, W. K. Woelk, M. A. Koroscil and K. T. M. Col, Bacterial contamination of paper currency, *South. Med. J.*, 2002, **95**, 12.

6. J. M. Maritz, S. A. Sullivan, R. J. Prill, E. Aksoy, P. Scheid and J. M. Carlton, Filthy lucre: A metagenomic pilot study of microbes found on circulating currency in New York City, *PLoS One*, 2017, **12**, 4.

7. Y. Zuo, K. Zhang, J. Wu, C. Rego and J. Fritz, An accurate and nondestructive GC method for determination of cocaine on US paper currency, *J. Sep. Sci.*, 2008, **31**(13), 2444–2450.

8. M. T. Sampson, New study: Up to 90% of U.S. paper money contains traces of cocaine, American Chemical Society, 2009, https://www.acs.org/content/acs/en/pressroom/ newsreleases/2009/august/new-study-up-to-90-percent-of-us-paper-money-contains-traces-of-cocaine.html.

9. Dublin City University, 100-percent contamination of euro notes with cocaine, ScienceDaily, 2007, www.sciencedaily. com/releases/2007/01/070116093548.htm.

10. BBC News, Cocaine traces 'on Spanish euros', 2006, http:// news.bbc.co.uk/2/hi/europe/6208877.stm.

11. F. T. Carson and V. Worthington, The wearing quality of currency paper. U.S. Dept. of Commerce, National Bureau of Standards. Research paper RP1390, *J. Res. Natl. Bur. Stand.*, 1941, **26**, 467–480.

12. A. Phillipp and J. Wolfarth, *U. S. Pat.*, 9004257, Sensor device in a bank note processing machine, 2015, http://patft.uspto. gov/netacgi/nph-Parser?Sect1=PTO2&Sect2=HITOFF&p=1& u=%2Fnetahtml%2FPTO%2Fsearch-bool.html&r=1&f=G&l= 50&co1=AND&d=PTXT&s1=9004257.PN.&OS=PN/9004257& RS=PN/9004257.

13. G. Terry, *Pigments, Paint, and Painting: A Practical Book for Practical Men*, E. & F.N. Spon, London, 1893, p. 104.

14. W. T. Brande and J. Cauvin, in *A Dictionary of Science, Literature, & Art*, Longman, Brown, Green, and Longmans, London, 1852, p. 223.
15. D. T. Vu, Characterization and aging study of currency ink and currency canine training aids using headspace SPME/GC–MS, *J. Forensic Sci.*, 2003, **48**(4), 754–770.
16. U.S. Currency: How money is made. Bureau of Engraving and Printing, U.S. Department of the Treasury, https://www.moneyfactory.gov/uscurrency/howmoneyismade.html.

CHAPTER 6

Counterfeiting

Counterfeiting any of the bills of credit. . .or altering or defacing any of the said bills with intention to change the value or denomination thereof—is punishable, for the first offense, by standing in the pillory three hours, and having the right ear nailed to the pillory and cut off, and by receiving thirty-nine lashes on the bare back, and by being branded with a red hot iron on the right cheek with the letter C, and on the left cheek with the letter M [for malefactor]; which brand to be at least one inch in length and three-quarter of an inch in breadth.[1]

1811 statute as reported in "The Office and Duty of a Justice of the Peace. . . according to the laws of North Carolina"

During World War II the Nazis printed millions of banknotes. We're not talking about German marks, but rather pound sterling notes, as well as a few American dollars. In what can arguably be called the most successful counterfeiting operation of all time, dubbed Operation Bernhard, the Nazis printed upwards of £300 million—a staggering amount equivalent to 15% of all circulating currency in Britain at the time, and worth over £3 billion today. Even more incredibly, this entire operation was housed in concentration camps and manned entirely by the Jewish and Polish prisoners detained within them.

The Chemistry of Money
By Brian Rohrig
© Brian Rohrig 2021
Published by the Royal Society of Chemistry, www.rsc.org

Under the supervision of SS Major Bernhard Krüger, the initial intent was to drop planeloads of money over Britain, devaluing currency and raising inflation to the point at which the economy would be so destabilized that the Brits could no longer fund the war. This original plan would be abandoned, however, as it became evident to the Nazis that it would be far more lucrative to use these counterfeit bills to finance their own war efforts, as their own currency was faltering—by 1943 one mark was only worth 1/40 of a pound, and was for the most part worthless outside of Germany and its conquered territories.

The higher-denomination banknotes issued by the Bank of England at the time were so unimaginative that they were practically begging to be forged. They had changed little over the past 100 years. They were only printed on one side and in one color, black. Their main security feature was a watermark. The Nazis would commandeer a paper plant and painstakingly replicate the rag paper used to print the banknotes, including the watermark. During the early part of the operation the paper would be shipped to the Sachsenhausen concentration camp, where the banknotes would be printed using state-of-the-art equipment. To maintain secrecy the base of operations would periodically change locations throughout the war. Meticulously engraved copper plates were prepared in Berlin, capable of producing startlingly accurate reproductions. The Nazis even figured out the pattern to the serial numbers. The counterfeits were so good they even fooled officials from the Bank of England (Figure 6.1).

The fakes weren't perfect, however. Minor details were off. Owing to its intricacy, the vignette of Britannia proved especially difficult to forge, proving to be the banknote's best security feature, although it was not intended as such. To simulate wear and tear the bills would be folded and refolded and intentionally soiled so as to look used. Pin holes would be poked in the bill to cover up imperfections, lending an air of authenticity while simulating yet another attribute of used currency; pinholes were a common sight in banknotes in those days—new stacks of bills would be pinned together at the bank to bind them together. Despite these flaws, the counterfeited bills were so good they could only be identified as fakes by those who knew exactly what they were looking for, and then only if they were inspected closely. Even though the bulk of the counterfeit bills never

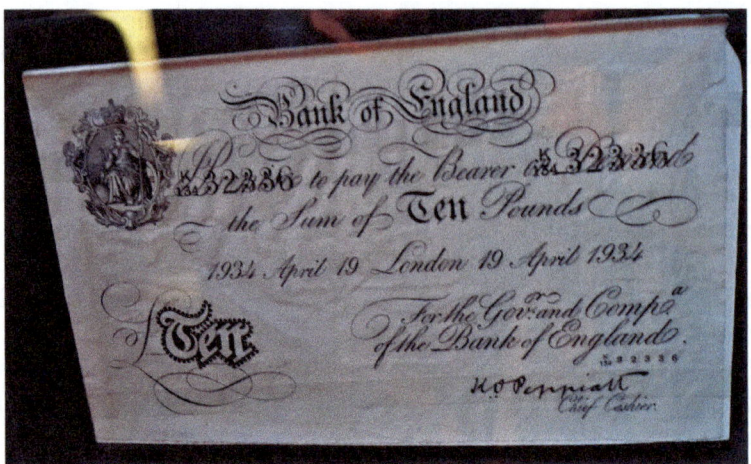

Figure 6.1   Forged Bank of England bank note from Operation Bernhard.
Image courtesy of Wikimedia/Joy of museums.

reached England's shores, officials were so troubled that they stopped issuing notes greater than £5 in 1943; eventually these notes would be reissued, but only after undergoing a thorough makeover. British authorities were actually apprised of the scheme by a spy in the early stages of the war—in response they issued a special wartime £1 blue and orange note with a metal thread running through it.

The Nazis had a large pool of talent from which to choose, enlisting those with the technical skills necessary to run a counterfeiting operation. In all, 142 of those detained within the camps would be recruited, and since it was a top-secret operation they lived and worked apart from the others. Although they were treated marginally better than others in the concentration camps, the plan was always for them to be executed at the conclusion of the project in order to maintain secrecy. Close to the war's end the order was given to execute all the workers who were involved in the operation, but the order explicitly stated they were to be executed all at once. The order was to be carried out at a separate facility, involving three separate trips to transport them. On the third trip the truck broke down and the workers had to endure a two-day march. By the time they arrived at the execution site, the Allies were at the door and the Nazis were preparing to flee. As they fled, they threw crates of the counterfeit bills into a nearby lake. The next day the camp would be liberated by the Americans, and the reluctant counterfeiters would live to tell of their extraordinary work. A number of the submerged bills would be recovered in 1959, and astonishingly, were still in good condition.

This episode served as a wake-up call not just for Britain but for the rest of the world. Even though counterfeiting was nothing new (after all, it has been dubbed the world's second oldest profession), still, the abolition of the gold standard and with it the replacement of coins with paper had forever changed the landscape of money. Paper and ink were easier to come by than furnaces and forges. Fake paper money could be churned out a lot faster than fake coins, and for the first time the specter was raised that a well-funded counterfeiting operation could decimate a country's economy virtually overnight. A sustained effort to thwart counterfeiting thus became the mandate for every country that produced banknotes. A banknote that could not be counterfeited became their holy grail.

6.1   PENALTIES FOR COUNTERFEITING

From its inception, paper money has been prone to being copied illegally. The very first banknote (the one-kuan note of 1375) contained the warning that "the counterfeiter shall summarily be decapitated". In the American colonies of the 18th century it was not uncommon to see the grim reminder "To counterfeit is DEATH" splashed boldly across the face of banknotes.

These written warnings were not just scare tactics. In England (and much of the rest of the world) counterfeiters were frequently put to death—typical punishments for men included hanging, while burning at the stake was reserved for women. The last woman to be burned at the stake in England was a counterfeiter, although she was actually strangled to death before burning, a common practice designed to show mercy—strangulation was considered a faster, more humane death than being burned alive.

In the early part of the 19th century, one in five executions in England and Wales was a result of forgery or counterfeiting, and one in three in London.[2] This alarming number of executions followed the issuance of the new £1 and £2 notes by the Bank of England in 1797, notes which were notoriously easy to counterfeit. The rise in public deaths owing to counterfeiting helped to turn the tide against capital punishment in England in the 19th century, resulting in the passing of a number of laws making it considerably more difficult to put someone to death. Most counterfeiters were not your average street thugs, who generated little sympathy, but rather respectable middle-class merchants and businessmen who didn't fit the stereotype of a hardened criminal; juries often found it hard to put these otherwise upstanding citizens to death.

The harsh punishment for counterfeiting in a way represented the first steps toward counterfeit prevention, but when one considers the pervasiveness of counterfeiting in the face of such draconian measures, the effectiveness of such a strategy can rightly be questioned. When paper money took on a life of its own, those who printed it quickly came to realize that unless they made it really difficult to duplicate, even the threat of death and dismemberment would not be enough to stop people from printing it themselves.

The average banknote today contains about 20 different security features, although there are closer to 50 different anticounterfeiting measures that have been used at some point on paper money. There are three levels of security within a banknote. The first level includes features that are designed to be seen by the public. Second level features are a little more challenging to find, and require the use of an instrument, such as a black light. Level three security features are machine readable using technology only available to central banks and are never revealed to the general public—these proprietary markers can be incorporated into the substrate, ink or varnish, and can only be revealed by sensors within banknote processing systems specifically programmed to detect them.

As merchants don't have access to high-tech toys that can authenticate currency with 100% certainty, counterfeiters can get away with passing funny money as long as their fakes are just good enough. They don't need to be able to fool the Federal Reserve, just Joe Schmo down at the corner deli. Most fake bills will eventually be discovered, as it's only a matter of time until a banknote will find its way to a central bank. But until that happens, counterfeit notes will circulate just like real money (Figure 6.2).

Figure 6.2   Printing Australian dollar notes.
© iStock/Getty Images Plus.

6.2   WATERMARKS: THE FIRST ANTICOUNTERFEITING MEASURE

The very first European banknote, issued by the Bank of Stockholm in 1661, contained several features designed to prevent unlawful duplication. It was printed on sturdy white paper and contained a lot of handwritten detail, a common sight even on the earliest banknotes. However, the most effective anticounterfeiting feature was the incorporation of a watermark spelling out the word "BANCO" on every note.

The ethereal watermark often goes unnoticed but is actually quite common. They are typically inserted into paper as a way to lend an identifying mark to one's work, or for security purposes. A true watermark will always be embedded into the paper itself. Digital "watermarks" inserted in documents today are not true watermarks. The term *watermark* is derived from the initial method used to make it, when the paper was still in its wet pulpy state (90% water and 10% pulp).

There are two ways to make watermarks. One way is to use an instrument known as a dandy roll, which consists of a wire screen bearing the image of the watermark placed upon a roller. The image is rolled onto the paper while it is still wet. As the image is imprinted into the pulpy mix, the fibers are spread apart in some areas while necessarily being compressed in others, forming the pattern of the watermark. Another method uses a cylinder mold, which is similar to relief printing—a raised area bearing the image of the watermark is impressed onto the wet paper, creating a 3-dimensional image. A cylinder mold is capable of creating finer detail than a dandy roll and is the preferred method for placing watermarks in banknote paper today.

Regardless as to how they are created, every watermark is essentially the same thing—an area with varying thicknesses placed within a piece of paper; as a thinner area allows more light to pass through than a thicker one, the embedded image can be revealed. The darker parts of a watermark are always denser than the lighter parts.

If a piece of paper contains a watermark, it can generally be revealed by holding it up to the light—if a watermark is present it should be visible. Just about any type of high-quality paper will contain the manufacturer's watermark. In the past, many postage stamps contained watermarks, and any serious stamp collector can tell you all about them. Only certain stamps contain watermarks, and a stamp with a watermark is usually worth more than the same stamp without one. As watermarks on stamps are usually difficult to see, a special type of fluid is required to reveal them.

When revealing a watermark, the philatelist will put the stamp in a black tray, and then add watermarking fluid. As the stamp gets wet, the transmission of light through the thinner areas of the watermark makes it visible; the black tray offers contrast, making the whiter areas stand out. The fluid dries very quickly and does not harm the stamp or the adhesive. The only purpose of the fluid is to get the stamp wet so the watermark can be seen. In the past, benzene was the fluid of choice for detecting watermarks, but as it is now recognized as a carcinogen, it is no longer used.

Commercially available watermark detecting fluids today are typically a mixture of aliphatic hydrocarbons, which lack a benzene ring; their carbon atoms are often arranged in a straight line. Benzene, on the other hand, is an aromatic hydrocarbon, containing a six-carbon ring with alternating double bonds, making it quite stable. Aromatic hydrocarbons are usually characterized by a strong odor, hence their name. Aliphatic hydrocarbons can also be arranged in a cyclic fashion, but they lack the alternating double bond structure of a benzene ring.

Both classes of hydrocarbons are volatile nonpolar compounds though, as it is essential that the fluid evaporate as quickly as possible so as not to damage the stamp. As nonpolar fluids have weak intermolecular bonds, they evaporate quickly. Water is not used to test for watermarks on stamps because it will dissolve the adhesive gum on the back, greatly decreasing the stamp's value. Nonpolar solvents, being insoluble in water themselves, will not harm the water-soluble gum.

Water reveals watermarks because paper, like fabric, tends to become somewhat transparent when wet. If you've ever worn a white T-shirt in a swimming pool you know how little coverage it gives you when wet. You may also have noticed that paper and fabrics tend to get darker when they get wet—this somewhat contradictory observation is the key to understanding why they also become more transparent. The darker an object, the less light it is reflecting. The water within the wet paper causes less incident light to be reflected and more to be absorbed, which can then be transmitted through the paper. In dry paper the internal spaces are filled with air, which has little effect on incoming light waves; as a result, most light that strikes dry paper is reflected, rendering it opaque.

The ability of a medium to refract, or bend light, is known as its index of refraction. A vacuum doesn't bend light at all, so its index of refraction is 1. Air has an index of refraction just slightly above 1, while that of water is around 1.33. The reason you see water in air is because it refracts light differently. If two objects have the same index of refraction, you will not be able to distinguish between them as they will both bend light at the exact same angle. Certain types of glass appear completely invisible in certain types of oil because their indices of refraction are the same. A classic magic trick is to shatter a drinking glass and drop the pieces in a container of vegetable oil. The magician says a few magic words and then, voila, he pulls out a completely whole drinking glass. The glass was actually there the entire time, but as its index of refraction is the same as that of the oil, it was invisible so long as it stayed immersed.

The index of refraction of cellulose is around 1.47, which is fairly close to that of water. When fabrics get wet the water fills in the spaces between the cellulose fibers; as both refract light similarly, the paper appears somewhat transparent. In essence, the water is directing the light waves to pass through the fabric, unlike air which ignores the light waves, enabling them to be reflected by the fabric.

To observe the watermark on a banknote you can employ the same method as that used for stamps, but you don't need to acquire any special fluid to do so, as water will not harm a banknote. The watermark can easily be seen if the note is wetted and placed on a black surface.

Curious as to what types of watermarks could be found on today's money, I reached into my wallet and plucked out a US $5 bill. After getting it wet, I laid it flat on the black surface of my kitchen stove. Sure enough, several watermarks instantly appeared—a large number "5" off to the right, and a vertical row of three "5s" to the left.

You can also reveal a watermark by holding the bill directly in front of a bright light. If you hold a $10 bill over a bright light, on the obverse side you can clearly see a watermark containing a smaller head of Alexander Hamilton off to the right of the main portrait. On the reverse side the image is reversed. A true watermark will always be visible from both sides of the bill, with one image being the reverse of the other.

Watermarks are ubiquitous in currency. Every euro banknote contains a watermark denoting the denomination. Watermarks denoting the value are quite common, discouraging a counterfeiter from changing the denomination of a banknote into one that is higher. Paper £20 banknotes contain a watermark of Queen Elizabeth II off to the left, which can be seen under a strong light; a much brighter £20 denotation will be visible as well.

Watermarks are a key feature on Japanese yen banknotes. On the 1000, 5000 and 10 000 yen a large blank oval is prominently displayed in the center of the bill, on both the face and the back. When held up to the light, the image of whomever is displayed on the bill appears in the oval (Figure 6.3). Other watermarks are also seen within the yen; depending on the denomination, one, two or three vertical bars appear when held up to the light or otherwise revealed. You can see blank ovals on the backs of some American banknotes as well, such as in the aforementioned $10 bill, which reveals the watermark when held up to the light.

Despite all of the other high-tech security devices available today, the watermark is still one of the most widely used security features, not just in banknotes, but in a host of other documents that are susceptible to forgery, such as birth certificates,

Figure 6.3 Japanese yen watermark on notes.
© iStock/Getty Images Plus.

passports and checks. At the time of writing, UK passports contain a watermark of William Shakespeare.

Watermarks are inexpensive to make and yet are nearly impossible to reproduce. On the forged Nazi banknotes from World War II, the watermark was more prominent than it was on the genuine notes, a defect that would be hard to notice unless you did a side-by-side comparison between a counterfeit and a genuine note. The proper placement of the watermark was also a challenge—as paper shrinks when it dries, a watermark initially has to be made slightly bigger, making it difficult to line it up precisely with certain features on the banknote.

6.3   BEN FRANKLIN: MASTER ANTI-COUNTERFEITER

If you look at a $100 bill, the highest US banknote in circulation, you will see the visage of Benjamin Franklin, the father of US currency. Although he was not the first to print money in America, he printed much more than anyone before him, and many of his innovations are still in use today. But Franklin did more than just print money—he also wrote extensively about the role of money in society and, ever the innovator, proposed alternatives to the current monetary systems then in place.

In Franklin's time, paper money was still a novelty, and its entire structure was different than what we experience today. The early forms of paper money were never legal tender in and of themselves—they were always backed up by gold or silver coins. Money was not printed by a central government but by individual states, who often delegated this authority to banks. Banknotes even had an expiration date! There were a lot more denominations than we have today. In 1779 there were at least 16 denominations being circulated in the American colonies—among them the $1, $2, $3, $4, $5, $20, $30, $35, $40, $45, $50, $55, $60, $65, $70, and $80 notes.

In addition to using the latest printing techniques to mass produce money, Franklin came up with all sorts of ways to thwart counterfeiters. He designed his bills with intricate patterns and lettering that would be hard to duplicate. He used several different types of fonts on each bill, making some of these early bills vaguely reminiscent of ransom notes. On one set of bills he intentionally misspelled the word "Pennsylvania," assuming that a counterfeiter would most likely spell it correctly. Franklin incorporated a variety of printing techniques, using both intaglio and relief printing.[3]

Along with his friend, naturalist Joseph Breintnall, Franklin devised a technique known as nature printing, whereby an impression of a leaf could be transferred to the metal plate of a printing press and then printed onto a banknote. He knew that the intricate design of a leaf would be extremely difficult to copy, and the subsequent paltry numbers of confiscated bills containing forged nature printing proved Franklin right.

To create a new design, all that was needed was a new leaf, as no two leaves are ever identical. Furthermore, as the veins on a leaf gradually taper off, becoming thinner and thinner until they terminate, it would be almost impossible for an engraver to replicate this kind of detail. The first appearance of nature printing was on a Pennsylvania 20-shilling note printed in 1739, with three blackberry leaves and a willow leaf. As far as we know, Franklin never revealed the technique behind this wholly unique method of design. It has been speculated that he began by first making a mold of a leaf in plaster, filling his mold with molten metal to make a casting. Intricate designs of leaves have been a mainstay on paper money ever since and still feature prominently on the US $1 bill (Figure 6.4).

Franklin was responsible for a number of innovations beyond design. He worked closely with paper mills, enabling him to experiment with the incorporation of various substances within the paper itself. He was the first to embed colored threads within currency paper, using a different colored thread for each denomination. He used green threads for 10-shilling bills, red in 15-shilling notes and for 20-shilling banknotes he used blue.[4] As we have seen, the practice of incorporating colored threads was later adopted by the Crane Company, and is still practiced today.

Franklin also incorporated tiny bits of mica and asbestos in the paper. Just like with other security features, it was only a matter of time before these innovations would be copied. Counterfeit bills from the 18th century have been confiscated that not only contain silk threads, but also bits of mica. Nonetheless, part of Franklin's genius was to incorporate numerous

Figure 6.4   A one (US) dollar bill.
© Shutterstock.

security details within a single banknote—he realized that it was the accumulation of many features at once that made a note secure.

Although the use of mica in banknotes did not extend very far past colonial times in America, Franklin would prove prescient once again in his use of a security feature, as today mica is a key component in color shifting inks, which we will examine later. Mica can be found today in mascara and metallic paints—if your goal is to make something glittery, mica is your best bet. It has a pearlescent luster, an almost indescribable way of reflecting light that can best be understood by examining a string of pearls. As it has such excellent cleavage, mica is often found in thin sheets, interacting with light in much the same way as would a soap bubble or film of oil. Mica is naturally iridescent, changing color when the angle at which light strikes it changes. Once again, we see thin-film interference at work—as light waves bounce off the various surfaces of the thin mica sheets, the reflected waves constructively and destructively interfere, producing iridescent effects.

6.4   INDENTED CURRENCY

Another innovation that appeared in early American money was that of indented currency. A series of printed notes would be bound together to form a pad. At one end of each note would be an elaborate design. When a note was issued, it would be cut in a wavy pattern through the center of the design. The issuing party would keep the stub, which also contained a matching serial number. To redeem your note, you would take it back to the issuer—both the serial number as well as the indented top of the note had to match that of the stub. It was an ingenious method and virtually foolproof, but eventually became impractical as the amount of money in circulation continued to grow.

6.5   MAGNETIC INK

If you dangle any US banknote between your thumb and forefinger and bring a strong neodymium magnet toward the bottom of the bill, it will be drawn to the magnet. If you really want to see the magnetic nature of currency ink, add some water and a banknote to a blender and press "Puree". In no time you will have a dollar-bill smoothie. If you swirl a magnet in the slurry, a good bit of it will stick to the magnet.

Magnetic ink is not made by dumping some iron filings in ink and then mixing it up. As iron is so much denser than ink, the iron particles would settle to the bottom unless continually agitated. To incorporate iron into ink, its particle size must be reduced to the point at which it will remain in a colloidal suspension. This feat is accomplished by making a ferrofluid—a mixture of iron oxide particles in suspension.

A ferrofluid can be made by combining a mixture of iron(II) chloride and iron(III) chloride in a basic solution, which will form magnetite (Fe_3O_4), as described by the following reaction:

$$FeCl_{2(aq)} + 2FeCl_{3(aq)} + 8NH_{3(aq)} + 4H_2O_{(l)} \rightarrow Fe_3O_{4(s)} + 8NH_4Cl_{(aq)}$$

A surfactant needs to be added to keep the particles from clumping together. In order to stay suspended the particles must be around 10 nm or so in diameter. This small particle size enables them to remain in suspension without settling out, as well as ensuring their even distribution in ink.

The black gunk that collects on a magnet dipped in shredded currency is a type of ferrofluid. A simple qualitative test can be used to determine if it contains iron. First, it must be dissolved in a little strong acid. Then a few drops of thiocyanate solution are added—the formation of a blood red color confirms the presence of iron, owing to the formation of the iron(III) thiocyanate complex ion ($FeSCN^{2+}$).

Magnetite is not ferromagnetic but paramagnetic, being only slightly attracted to a magnet, which explains why magnetic currency is not attracted to an ordinary household magnet. Although iron itself is ferromagnetic, many iron compounds are only paramagnetic. The paramagnetic nature of iron compounds can be demonstrated by attaching an iron tablet (containing iron(II) sulfate) to a strand of dental floss and then slowly bringing a magnet towards the tablet—the tablet will be readily drawn to the magnet.

Ferrofluids were originally developed by NASA for use in the space program. Even though the US space program has received a lot of flak from the public for spending billions of dollars to put men in space, the benefits of the program to everyday citizens cannot be discounted. Ferrofluids provide a great example of one of these lateral benefits, as they have a variety of high tech applications. As disk drives and X-ray generators rotate at such high speeds, they make use of ferrofluids to reduce friction and vibrations. Ferrofluids have been used to dissipate excess heat from loudspeakers, as well as serving as a contrast agent in magnetic resonance imaging.

Ferrofluids are fascinating to look at—if a magnet is brought close to a ferrofluid that is encased in glass, handsome spikes are produced that can be elongated and distorted by adjusting the position of the magnet. Ferrofluid displays are popular in science centers and children's museums.

Magnetic ink is a good example of a level two security feature. As most people do not have a neodymium magnet on hand, they are unable to test for this feature themselves—by making security features that aren't too easy to discover, legitimate moneymakers are banking on counterfeiters leaving something out, providing an easy way to identify a fake.

As the metal in magnetic banknotes can be detected with a good metal detector, the US government has toyed with the idea of using metal detectors to determine if someone is carrying large sums of cash. Privacy advocates though fear that such a course of action would be overly intrusive, as the authorities might someday be able to tell exactly how much money you are carrying just by scanning you with a metal detector. Although initial studies have shown that such an approach is feasible, these fears are somewhat unfounded, as the technology is not sufficient to determine the actual value of the money you might have in your possession. As banknotes of different denominations each contain roughly the same amount of magnetic ink— a \$100 banknote does not contain any more metal than a \$1 note—the best an ordinary metal detector can do is perhaps determine the number of magnetic bills you are carrying, but not how much they are worth.

Vending machines that accept banknotes must be able to discern one type of note from another, however, as do machines that give change for a dollar, or five dollars. The older types of machines relied on the same technology that powered the cassette tape—magnetic recording heads that rely on induction. A cassette tape produces sound owing to tiny iron oxide particles arranged in a specific pattern on the tape. When the tape passes by the magnetic head of the cassette player, which is a tiny electromagnet, an electric current is induced in the head, creating a signal which is converted into sound after it flows through the little wire wrapped around it.

When a cassette tape is played, moving magnetic fields give rise to electrical fields; in the same way, as the magnetic ink of the banknote rubs against the magnetic heads of the money changer, the presence of magnetic material induces an electric current in the heads, which is then converted into a signal that tells the machine whether or not to accept the bill. If a counterfeit bill is inserted that does not contain magnetic ink, that bill will be summarily rejected. However, if magnetic ink is used, the bill still might not be accepted; it is not enough that magnetic ink simply be present, but rather it is the specific pattern of magnetic ink that matters, a pattern that differs from one denomination to another. Each banknote has its own magnetic fingerprint, which is only readable by a sensor specifically programmed to detect it. If you have a banknote that is rejected by a machine it might be time to take a closer look at it—it just might be a fake.

Although at one time magnetic ink was a novelty, today it is obtained quite easily. Most checks printed today use magnetic ink—banks use special scanners to detect it. The more popular types of check scanners utilize Magnetic Ink Character Recognition Technology (MICR), which reads the magnetic ink used to print the specialized characters at the bottom of a check that contain the routing and account numbers. Laser printers and copiers often use magnetic toner as well. Security features originally used only for currency often end up finding their way into the broader market, forcing the makers of banknotes to come up with ever more ingenious methods to protect their products.

Although magnetic ink is a common feature in US dollars, it is by no means exclusive to American currency. Euro banknotes, as well as a host of others, also contain magnetic properties. Serial numbers are commonly printed using magnetic ink. Sometimes the amount of ink used in a banknote is too slight to be detected by a magnet, but if you are determined to test your currency for magnetism you can always purchase a magnetic currency detector online relatively cheaply; the low-end models, however, don't always work as well as advertised, sometimes giving false readings.

6.6   ULTRAVIOLET FEATURES

Today's money contains many of the same security features as the earliest banknotes. The substrate, and the features incorporated therein, are still of utmost importance, as are the elaborate, and increasingly colorful designs. Seldom is a security feature discarded, but rather new ones are just tacked on. If you look at a typical banknote today there is so much going on you don't know where to look first. One relatively recent, and extremely effective, innovation is that of using fluorescent markings. Even though ultraviolet light was discovered in 1801, it was not prominently used in currency until late in the 20th century.

Whereas blacklights used to be seen as more of a counter-cultural thing, in recent years they have made a resurgence. You can walk into just about any department store and purchase a blacklight, along with a plethora of products that fluoresce underneath it. You can purchase blacklight sensitive posters, candles, paint, markers, bubbles, clothing, nail polish, hair coloring, and jewelry. If you turn off the lights in your home and illuminate the surroundings with a blacklight, another world will be revealed. Any number of dishes, labels, fabrics, toys and wall hangings will fluoresce. Every scorpion will glow bright green under the blacklight owing to a fluorescent pigment in the exoskeleton. The next time you consume an energy drink, bring out the backlight—it will fluoresce brilliantly, as will tonic water (owing to quinine). Forensic examiners use blacklights to test for bodily fluids—only urine and semen fluoresce unassisted. For security purposes you can buy invisible marking pens that write with ink that is only visible under ultraviolet light. If you illuminate your credit card, driver's license or passport under a black light there is a good chance you will see an image you did not see with the naked eye.

Ultraviolet light is roughly divided into three types: UVA, UVB and UVC. UVA has the longest wavelength and is typically used in blacklights; there are, however, long and short wave blacklights—if identifying fluorescent minerals, you will need to use both types, as some minerals only fluoresce under shorter wavelengths. We get a suntan by exposing ourselves to the shorter wave UVB light; fortunately, UVC is completely blocked by the ozone layer.

This energy transformation from ultraviolet to visible is responsible for fluorescence. The energy transformation occurs because energy is quantized, meaning it only exists in discrete little lumps, or quanta. This principle forms the basis of quantum mechanics. Each energy level in an atom corresponds to a specific amount of energy, so there is a certain amount of energy that an electron must absorb to reach a certain energy level. As the electron returns to the ground state, the specific color of the light emitted depends on the amount of energy that is released. Colors near the violet part of the spectrum carry more energy than those nearer the red part of the spectrum.

Fluorescence occurs when one type of electromagnetic radiation is absorbed and another type is emitted, usually of a longer wavelength, such as when a pigment absorbs UV light and emits visible light. A fluorescent bulb is so named because the white phosphor coating on the inside of the bulb converts ultraviolet radiation into visible light. During this process, an electron within a fluorescent pigment absorbs energy and becomes excited, jumping up to a higher energy level. However, the excited state is only temporary, and the electron soon falls back down to the ground state—as it does so the previously absorbed UV energy is released as visible light.

If a banknote looks different under a black light than it does under visible light, it most likely contains a fluorescent feature. With the notable exception of the US $1 bill, most banknotes will exhibit some degree of fluorescence. In my stack of a hundred banknotes from around the world, nearly every one fluoresced under the blacklight. It is common to see serial numbers printed in bright orange fluorescent ink. You can find an assortment of fluorescent patches, strips and designs. The entirety of some Iraqi banknotes will fluoresce a bright blue under the blacklight, as the paper itself is fluorescent.

Beginning with the $5 denomination, every US banknote contains an embedded fluorescent polyester security thread running through its width. It is woven through the paper, alternately appearing on the face and back of the bill, making it possible to remove the thread without ripping the bill in half, a course of action not recommended, though, as doing so will cause the bill to fail a crucial test of authenticity. These threads first appeared in 1990 on $100 bills and by 1993 were incorporated into other denominations. Under a blacklight, each thread appears as a single continuous strip, with each denomination fluorescing a different color. The thread in the $5 bill appears blue, in the $10 bill it looks red and in the $20 note a green fluorescence can be observed. The $50 thread is yellow, and the $100 bill strip fluoresces pink. The $100 bill also contains a wider blue security thread to the right of the portrait. In normal light these threads are all but invisible, but glow vividly under the blacklight. The microprinting on the thread, which reveals the denomination, is invisible under ordinary white light, but is clearly visible under ultraviolet light.

If you put an Australian $5 note under the blacklight you will see a bright green bird—the eastern spinebill. The $10 note features a fluorescent cockatoo, while a luminescent swan appears on the $50 bill. The next time you encounter Chinese money, look for fluorescent dragons. All euro banknotes (€5, €10, €20, €50, €100, €200 and €500) contain the same fluorescent features—on the face of the note is a green fluorescent flag and orange stars. On the back you can see a bridge as well as the number representing the denomination—both of these will glow brightly under the UVA radiation emitted by a blacklight. Under UVC radiation, however, the fluorescent features will fluoresce a different color. Embedded within the paper are tiny little fibers that fluoresce a bright red, blue or green under the blacklight—these are put in place during the production of the security paper itself, while fluorescent features on the surface are made by incorporating special pigments within the ink.

Although most banknotes will fluoresce under a blacklight, you are not likely to see one glow in the dark. Fluorescence is often confused with phosphorescence. Although all objects that glow in the dark are fluorescent, not all fluorescent objects glow in the dark. In a fluorescent object, light emission ceases as soon as the excitation source is removed. If excitation continues for more than a millisecond after the illuminating source is removed then phosphorescence has occurred, commonly referred to as "glowing in the dark".

UV counterfeit detectors are becoming a common sight in retail establishments. When a banknote is inserted into the device a blacklight makes the fluorescent markings clearly visible. Any ordinary blacklight will accomplish the same purpose.

6.7 OPTICAL BRIGHTENERS

Although many of the security features added to currency are highly fluorescent under a blacklight, the paper itself is decidedly non-fluorescent (usually). If you put most types of plain white paper under a blacklight you will see some degree of fluorescence, but not on the cotton-linen paper used for banknotes—it is dull and drab by comparison. The substrate in polymer notes is typically not fluorescent either. Most types of paper contain optical brighteners, which are added to make paper appear whiter—these whiteners are intentionally left out of the paper used to make banknotes. Paper with no optical brighteners is referred to as optically dead, and is often preferred by artists. As optical brighteners are degraded by atmospheric oxidation, the appearance of the paper changes over time, plus, paper that contains these whiteners looks different when viewed under different types of light.

Optical brighteners are typically added to ordinary paper during the manufacturing process. If added while the pulp is wet they produce internal whitening; for more surface whiteness they can be added while the paper is drying, tending to appear more bluish.

One of the first optical brighteners was made from the bark of the horse chestnut tree (*Aesculus hippocastanum*) which contains the substance aesculin, a naturally fluorescent compound that fluoresces bright blue in solution (Figure 6.5). Today most optical whiteners are synthesized in the lab, usually involving a derivative of the cyclic compound stilbene, a complex organic substance known for its blue fluorescence.

Optical brighteners are added to laundry detergent and textiles, making clothes appear "whiter than white", as under sunlight they absorb invisible UV radiation and reemit visible light, making clothes appear brighter. They are often added to cover up yellowish or off-color hues that may arise during the manufacturing process. As almost all paper will have optical brighteners added to it, the lack of these brighteners is one more way to make the job of counterfeiters more difficult.

If you want to get genuine currency to fluoresce it's not that difficult. If money goes through the wash, the banknote will not only be laundered, but will end up being fluorescent as well

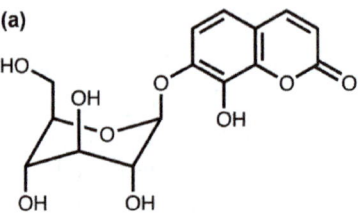

Figure 6.5 (a) The compound aesculin from the horse chestnut tree. (b) Horse chestnut (Aesculus) fruits with leaves and flowers. © Shutterstock.

(assuming detergent was used). The optical brighteners used in detergent have a high affinity for cellulose, and thus bind tightly to banknotes containing these fibers. Optical brighteners especially love cotton. If you ever find yourself being falsely accused of passing counterfeit bills because of the fluorescent paper, try to remember if they ever went through the wash. These optical whiteners can be removed by placing your banknotes out in the sun for a day or two—they so effectively absorb ultraviolet radiation that exposure to such will degrade them in a short time. Or you can try repeatedly rinsing in water, as most optical brighteners used in laundry detergent are water soluble.

6.8 FLUORESCENCE LIFETIMES

UV fluorescence is not the only type of fluorescence that can be used to analyze banknotes; as fluorescence occurs anytime an object absorbs, and then emits, a different type of electro-magnetic radiation, the excitation source can come from a wide range of the spectrum—X-rays or even visible light can be used to excite electrons. Typically, the emitted photons are of a lower energy than the absorbed photons. Therefore, even though currency paper does not fluoresce under a black light, it can be made to fluoresce when exposed to certain high energy lasers and the like, in which case it will absorb and emit photons of different wavelengths, enabling its fluorescence spectrum to be determined. If a banknote of the same denomination is tested and is found to exhibit a different spectrum than that of a known genuine note, then that note is suspect.

There are any number of textiles and/or paper products that might exhibit a similar fluorescence spectrum to currency paper, however, so fluorescence alone is usually not enough to enable one to say for certain that a note is genuine or counterfeit. However, if one examines the fluorescence lifetimes of the emitted radiation, a whole different story emerges. The fluorescence lifetime refers to how long a molecule exists in the excited state before falling back to the ground state and emitting a photon. Fluorescence lifetimes can vary significantly from one molecule to the next, but most are very short, ranging from a few hundred picoseconds to just a few nanoseconds. The longer it takes for a molecule to fall back to the ground state, the longer the fluorescence lifetime. A phosphorescent substance, for example, would have a very long fluorescence lifetime. However, an object does not have to glow in the dark to have a measurable fluorescence lifetime.

Using some pretty fancy high-tech equipment, researchers from the Department of Biomedical Engineering at Yale University performed a study to see if they could distinguish between counterfeit and genuine US $100 bills by measuring their fluorescence lifetimes. These tests are not ones you can perform in your basement, or even in the typical lab. To gain insight into the sophistication of their experimental procedure, consider the following description of the apparatus used:

> We used a custom-built two-photon microscope based on an Olympus BX51 WI upright fluorescence microscope. The excitation source was an 80 MHz pulsed Ti : Sapphire (Ti : S) laser tunable between 710 nm and 990 nm. The excitation wavelength was set to 735 nm with a 100 fs pulsewidth. The microscope objective was a 4×, 0.28 NA air objective.

Samples were held flat on a motorized 3-axis microscope stage Power at the sample was ~15 mW. The intrinsic fluorescence spectra of samples were obtained using a fiber optic spectrometer connected to the two-photon micro-scope. Fluorescence generated at the sample was reflected by a dichroic mirror and focused with a plano-convex lens into a 1 mm core fiber optic to maximize collection effi-ciency. Fluorescence lifetime capabilities were made pos-sible through the addition of a multi-channel plate PMT (photomultiplier) and a time-correlated single photon counting card. The fluorescence of the paper money was filtered through a 555 nm short-pass filter. (Excerpt repro-duced from ref. 5 with permission from The Optical Society, Copyright 2009.)

The researchers found that when excited by laser pulses with a wavelength of 735 nm, the fluorescence spectrum of currency paper ranged from 400–650 nm, with the 460–640 nm range being the most dominant. Counterfeit bills exhibited distinctly different fluorescence spectra, with significantly shorter fluorescence life-times than those of genuine currency. It was also observed that counterfeit banknotes made by bleaching the ink from a genuine note and then printing over it had significantly lower fluorescence lifetimes. This lower fluorescence lifetime was probably a result of a change in the paper's properties from exposure to the chemicals used, or else from residual chemicals that remained in the paper, altering its fluorescence lifetime signature.[5]

The use of ultraviolet radiation to illuminate fluorescent fea-tures on banknotes is so well known that it's easy to overlook its invisible cousin—infrared radiation (IR), which can be just as useful. To "see" the IR features on a banknote, you must view it with an IR sensitive camera. If a euro is placed under infrared radiation in an otherwise darkened room, you won't see a thing with the naked eye. But if viewed with an infrared camera you will see quite a bit—the denomination and some architectural images will be clearly visible but the rest of it will have seemingly disappeared—it looks as though the bill is incomplete, as though only partially printed. The regions of the banknote that appear white are reflecting IR radiation, while the parts that appear black are absorbing it (Figure 6.6). This feat is

Figure 6.6 Checking a 50 euro paper note to determine if it is counterfeit or genuine using infrared light.
© Shutterstock.

accomplished by using different types of ink on the banknote—if the ink reflects IR radiation it will appear white, if it absorbs IR it will be black.[6]

As we've all seen blacklights and tanning beds, ultraviolet radiation is typically better understood than is infrared, which is a bit more of a mystery. I first became enamored with infrared radiation while witnessing a demonstration at a physics teachers conference. The presenter grabbed a frisbee and set it on a table. He then recorded it with an IR camera and displayed the resulting image—the portion of the frisbee that had been grasped glowed brightly, forming a perfect outline of the presenter's fingers. It was an elegantly simple, yet incredibly powerful demonstration of infrared radiation.

Infrared radiation is usually invoked when discussing heat, as in thermal imaging or the radiation of body heat. Although it is true that every object above absolute zero is radiating infrared radiation, to say that infrared radiation is the same thing as heat is misleading. Heat always involves the transfer of energy from hot to cold, or, if you prefer, from a region of high thermal energy to one of low thermal energy. If I touch the coils of a hot stove my hand gets burned—the molecules in the stove are moving faster than the molecules in my hand. While it is true that the stove is emitting infrared radiation, most of the energy that goes into my hand entered *via* conduction.

If I was to hold my hand above the stove, I would primarily be feeling the effects of convection. However, if the stove was operating in a vacuum and I put my hand above the coils, then I would be feeling the effects of infrared radiation. It would feel hot because my skin absorbs IR radiation better than any other form, readily converting it into thermal energy, which my body senses as heat. Any type of electromagnetic radiation can produce heat, even intense visible light.

Even though infrared radiation is responsible for the heat given off by heat lamps, there are plenty of applications for infrared radiation besides heating. If you use a smartphone to take a picture of someone aiming the television remote control at you while activated, you will see a purplish dot on the picture. Smartphones are sensitive to IR and can pick up the infrared radiation emitted by the remote as it communicates with your television.

Infrared radiation involves energy at a very specific range of wavelengths, from about 700–1 000 000 nm (1 mm). Our eyes are sensitive to only the narrow band of radiation from 400–700 nm, so infrared is just out of our range. Just as there are different types of UV, there are different types of IR as well. Near infrared radiation is found at the very end of the IR band, and represents those wavelengths closest to visible light. Far-infrared radiation utilizes frequencies that are the farthest away from visible light. The types of frequencies involved in analyzing banknotes often fall in the near-infrared region.

Just as certain types of pigments are fluorescent under UV radiation, they can also be fluorescent under IR radiation. However, unlike UV fluorescence, in which the emission products are often in the visible light range, the emission products of IR fluorescence are typically not visible to the naked eye, so viewing it is a bit more challenging.

Every type of IR fluorescent ink will have both an absorption and an emission frequency. One type of IR security ink that is often used has an absorption frequency of 793 nm and an emission frequency of 840 nm. However, as both of these frequencies are in the IR region, and thus invisible to us, you can never see this ink with the naked eye—it is truly invisible ink.

Different methods have been developed that make it possible to "see" IR fluorescent ink. If a particular type of ink absorbs IR radiation in one wavelength and emits it in another, using a filter to block out the frequencies that are emitted will result in the ink appearing black if viewed with an infrared camera, as you are witnessing the absorbance of IR radiation. However, if you view the ink with a filter that blocks the absorbance frequencies it will appear lighter when illuminated, as you are witnessing the emitted wavelengths—the resulting image looks just like one produced *via* UV fluorescence.

The more sophisticated vending machines and bill changers rely on IR technology. The American $5 bill, for example, is commonly used in vending machines—it contains two narrow bands printed with IR sensitive ink. The $20 bill has a single narrow band, and the $100 banknote has a wide and a narrow band. Infrared sensors within these devices read this infrared signature—if the correct infrared pattern is not present the bill will be rejected; if it is present the bill is only accepted if it passes a myriad of other tests—no machine relies on one feature alone to authenticate a banknote.

6.9 FTIR SPECTROSCOPY

Using infrared radiation to detect counterfeit money extends beyond just viewing banknotes with an IR sensitive device. Different substances absorb radiation at different rates. You don't venture outside on a hot day in the summertime wearing black clothing unless you plan on sweating profusely. However, color is only one factor that determines how well a material will absorb a specific wavelength of radiation. The resonant frequency of the molecule also plays a crucial role.

Resonant frequency refers to the frequency at which something will vibrate with the least energy—it is often referred to as natural frequency. If the frequency of radiation corresponds to the natural frequency of a molecule, that particular frequency is likely to be absorbed. Carbon dioxide is an especially good absorber of infrared radiation, hence its importance as a greenhouse gas.

One method that analyzes the absorption of IR radiation is known as Fourier-transform infrared spectroscopy (FTIR). A FTIR spectrometer will produce an absorbance spectrum that will provide a unique identifying marker for any substance, and thus provide a means of comparison between the genuine article and a potential counterfeit. Each type of currency paper has a specific IR absorbance pattern that will differ significantly from that of a counterfeit. The ink can also be analyzed—each type used in printing a banknote has an IR absorbance spectrum that is all its own; so even though to the naked eye a counterfeiter's ink may look identical to official banknote ink, their IR fingerprint will be vastly different. The meticulous forger using a laser printer might do an exceptional job of reproducing a myriad of fine details, but the polymer base of printing toner will absorb IR radiation at a completely different rate than that of genuine currency ink, which can be clearly revealed by examining the FTIR absorbance data of both.

6.10   DIFFRACTION GRATINGS

Whereas UV and IR features involve the use of alternate light sources, there are a number of security features designed to be seen with the naked eye. These features utilize ordinary visible light. Falling under the wide umbrella term of optically variable devices (OVDs), these features are responsible for creating a dizzying array of special effects. OVDs have the ability to take white light and manipulate it in ways that stretch the bounds of credulity. Color-changing inks, holograms and simulated motion have all become as common as serial numbers on today's banknotes. There are literally dozens of different types of OVDs—some are made by using special inks, while others are incorporated into the substrate itself; still others are glued to the bill after everything else has been printed. The major players in security printing have devoted extensive resources into developing ever more elaborate measures to protect documents from forgery. However, every security device has a limited life span, as it's only a matter of time before counterfeiters will figure out how to replicate it.

One of the earliest OVDs appeared on the $10-dollar Australian polymer banknote, involving a revolutionary color-changing feature. First introduced in 1988, the banknote displayed an image of Captain Cook that changed colors when tilted. Even though color shifting devices are now fairly standard in all banknotes, they made quite a splash when first introduced. This particular optically variable device (OVD) contained Captain

Cook's image within a clear oval window, enabling it to be viewed from both sides. Sandwiched in the middle of the image was an extremely thin film of aluminum, which acted as the reflective surface. When tilted, a rainbow effect was created, causing Captain Cook to change colors.[7]

The best way to understand how this particular OVD works is to compare it to a compact disc, which also produces a rainbow effect when tilted a certain way. A CD is composed of a clear polycarbonate substrate (which is why you can see through a stack of CDs when viewed from the side) upon which is deposited a thin layer of aluminum, atop which is a clear protective acrylic layer. A CD contains millions of little ovoid microscopic pits—some are almost spherical and others are more paramecium-shaped, creating high and low areas (known as lands and pits). These microscopic bumps are arranged in spiral grooves that are about 1 μm apart, with thin layers of shiny aluminum in between the rows of pits.

When a light wave hits one of these reflective aluminum layers, it bounces off. At the same time numerous other light waves are bouncing off nearby slivers of aluminum as well, resulting in a lot of light waves being reflected at the same time. If the crest of a reflected light ray lines up with the crest of another reflected light ray, they constructively interfere; if the crest of one wave lines up with the trough of another, they destructively interfere. Different wavelengths are diffracted differently—the longer the wave the more it diffracts. Red waves will thus experience more diffraction than blue waves. This pattern of interference manifests itself as a rainbow of sorts.

The angle at which the CD is held will determine the pattern that forms, as the angle of incident light will determine whether a land or a pit is struck. If a light ray strikes a pit it may not be reflected at all; if it is reflected, it may emerge at a different angle than one striking a land. Light rays reflected at different angles will therefore produce different interference patterns, and thus different colors of light. By way of an analogy, suppose you were bouncing a tennis ball off the floor so it hit at a certain point on a wall you were facing. If there was a depression in the floor and the ball landed in it, the ball would rebound differently, maybe getting stuck in the hole or if it did make it out would most likely hit the wall at a different spot.

When a compact disc makes a rainbow, it is acting as a diffraction grating. If you have ever used a pair of "rainbow glasses", the paper-framed low-cost spectacles that act as a prism when they encounter white light, then you have seen a diffraction grating at work. Diffraction gratings represent an efficient, low-cost method to create a variety of optical effects, from creating rainbows to simulating motion. A diffraction grating is prepared by making a series of parallel striations very close to one another on the surface of a medium, thus creating row after row of very tiny slits. As light passes through these slits, it diffracts, or bends; in refraction light bends in one direction only, but during diffraction it bends in all directions, spreading out. As these waves interact with one another after they pass through the slits, an interference pattern is produced, forming all the colors of the rainbow.

The procedure used to create the color-changing image of Captain Cook was detailed in a US patent application:

> The 1988 plastic Australian ten-dollar note employs as an anti-counterfeiting security device, a diffraction grating image of Captain Cook which comprises a regular matrix of pixels each containing a curvilinear segment of one or more of the grating lines. The Captain cook pixellated diffraction grating image is substantially more secure than the traditional metal foil insert... It is able to sustain an acceptable level of structural stability as the notes become heavily crinkled in day-to-day use... The practical process effectively entails converting any given photograph of a portrait or scene into a set of data files that are used to control the exposure and writing characteristics of the electron beam lithography system in such a way as to enable the machine to produce corresponding master gratings according to the invention... First the portrait is scanned into a computer graphics system via a high quality colour scanner such as a Sharp JX-300 and then processed by a special purpose interactive software package... A durable metal master of the optimised grating may be obtained by vacuum coating the photoresist master with 200 Angstroms of 99.99% gold and electro-depositing a thick layer of nickel to act as a support... After separating from the glass master this gold coated nickel master may be bonded to a brass block and

used as a die for hot pressing of plastic film replica gratings… After metallising with aluminium and plastic coated for protection, the plastic replicas may be adhesively attached to currency notes or credit cards.[8]

Kinegrams represent one type of diffractive technology commonly used in banknotes today. They will usually be incorporated as foil stripes or patches, providing dazzling color-changing effects when tilted. On the £10 polymer note issued in 2016 by the Bank of England, featuring Jane Austen, the image of Winchester Cathedral is printed over a large translucent window in the center of the bill. The cathedral is a metallic gold on the face and a metallic silver on the back. If you tilt the bill slightly a rainbow effect is produced over this image as different colors move across it. On Australian banknotes kinegrams are used to produce a variety of effects, including birds that not only change color but also flap their wings when the bill is tilted. Kinegrams can also be found on Canadian dollars, Swiss francs, euros, and a host of other banknotes.

6.11 COLOR SHIFTING INK

Diffraction gratings are not the only way to produce a color change when the angle of illumination changes. If you examine a €50 note you can see the "50" change from purple to green when tilted. All American banknotes from $10 and up contain this same feature—the number of the denomination in the lower righthand corner shifts from copper to green when tilted. On the $100 bill a bell within an inkwell also changes color from copper to green when turned slightly, making it seem to appear and disappear as the inkwell itself is copper-colored. These features are not optical illusions—they are produced using optically variable inks (OVIs), which are applied just like any other ink. OVIs are used almost exclusively for security purposes. However, unlike diffraction gratings, which break white light up into its constituent wavelengths, OVIs produce their effects using thin-film interference.

The color of ink is always a result of the reflection of visible wavelengths of that color from the money's surface. Red ink is red because it is reflecting red light. The color that is reflected back does not usually depend on the angle at which light strikes it. However, just as in other optically variable devices, sometimes the angle at which light initially falls does matter. The angle at which light strikes a surface is known as the angle of incidence and refers to how much an incoming ray of light deviates from a ray that is completely perpendicular. An incoming ray of light that is perpendicular would have an angle of incidence of 0°. If an incoming ray strikes at an angle, the measurement of this angle from the perpendicular is the angle of incidence.

The law of reflection states that a ray must reflect off a surface at the same angle at which it struck a surface, so if the angle of incidence is 20°, then the angle of reflection must also be 20°. However, if you recall from our previous discussion of thin film interference, thin films can wreak havoc on this whole process, hence the name. If you have several very thin layers piled atop one another, then some light will be reflected from the top layer at the same time that some light will be reflected from the next layer, and so on, resulting in a bunch of reflecting light waves simultaneously emerging and interfering with one another, producing a myriad of possible color combinations.

Optically variable inks are made by suspending multilayered pigment flakes in normal ink. Each layer is produced as a result of vapor deposition, with successive layers deposited on top of one another. Each flake looks like two flakes that have been stuck together—layers on each side of the core are the same—effectively doubling the light-reflecting surface area, making the orientation of the flake within the ink less important. The base layer must not only be highly reflective but also be thick enough to be completely opaque—aluminum, a common choice, becomes completely opaque at 40 nm. A paint and coatings industry journal provides the following description of these optically variable pigments (OVPs):

> OVPs are multi-layer films of semi-transparent reflector (metals) and dielectric materials. These can be made of three layers, or a second type, which has a core of reflective material which in turn is covered by a symmetrical system of weakly refractive and partially reflective layers, for a total of five layers.
>
> In this second type of OVP, the center "inner reflector" layer is usually aluminum, and the "low refractive" layers can be either SiO_2 or MgF_2 or Fe_2O_3; and the "partially

reflective" outer layers can be Al, Cr, MoS_2, or Fe_2O_3. None of these layers has any color of its own, but when they are put into close contact with each other, they give rise to brilliant colors because of the components selective destructive and constructive interference as light reflects off the interfaces.

Changing the thickness of the interference layers changes the observed colors. The usual shifts are magenta to green; green to blue; and gold to green. Usually they are printed with intaglio presses with an engraving depth of 50–70 microns with a regular pattern. The pigments are usually 1 micron in total thickness and the tolerances are very tightly controlled. (Excerpt reproduced from ref. 9 with permission.)

Although there are a plethora of optically variable inks and other devices plastered all over our money today, they often exploit similar principles. As you may recall from an earlier discussion on the toning of coins, thin film interference is responsible for all the pretty colors you might see on a toned coin. On some of these coins, if tilted just so, the colors of the toning change because the interference pattern changes as the angle of incident light (and therefore reflected light) changes (Figure 6.7).

6.12 HOLOGRAMS

One of the more exciting security devices on banknotes today is the hologram. A lot of OVDs look like they are producing holograms, and may be mistakenly referred to as such, but just because a feature appears to be three-dimensional doesn't make it a hologram. You can find plenty of 3-D features on banknotes that aren't truly holographic. Although they were only introduced a relatively short time ago, the use of the hologram has experienced explosive growth in the security industry. A brief history of the hologram's use on currency is provided by ITW Technology, an industry leader in holographic technology:

The first banknotes to introduce holograms were issued in 1988 by Austria and Australia ... These two notes then paved the way for future success starting with Kuwait and Poland also adding patch holograms, Finland adding a

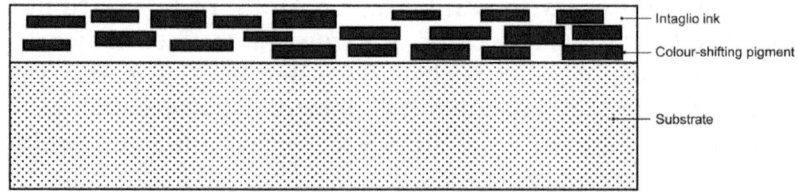

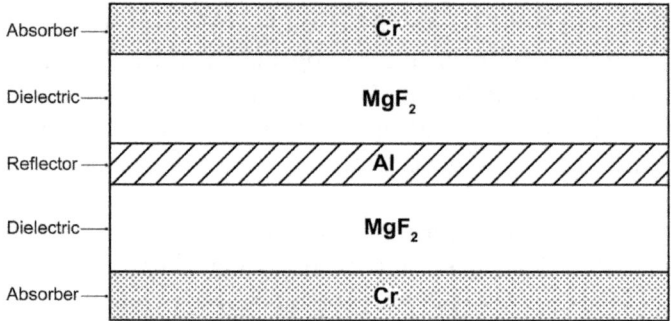

Figure 6.7   Cross section of color shifting ink and ink pigment.

thread in 1992 and Bulgaria adding the first ever holographic stripes on their 2000 Leva note in 1994. By the end
of 2007 patch and stripe holograms accounted for more
than 80% of all holographic features on banknotes with
more than 90 currencies featuring them on one or more
circulating denominations. It was further estimated in 2007
that 42 billion of the 125 billion notes produced featured a
hologram. As developments in the market continued and
polymer banknotes were introduced, the use of holograms
continued.

The first countries to launch polymer notes were Canada
and New Zealand and they incorporated patch holograms.
The UK quickly followed with their new £5 polymer note
with a holographic stripe. In the non-polymer arena, the
second-generation Euro banknotes (€20 and €50) included
holographic stripes and Israel issued new 50 and 200 Shekel
notes with single colour volume holographic stripes.
Switzerland also issued a new 50 Franc note with a holographic volume foil stripe and Poland's commemorative

20 Zloty banknote featured a colour shifting holographic feature similar to the ones on the Philippines 500 and 100 Piso notes. By the end of 2015 a total of 280 different denominations globally were using holograms, 132 used a patch hologram, 122 a stripe hologram and 26 a holographic thread.

To further demonstrate the importance of holograms in banknotes it is estimated that the average person only ever recognises two security features within banknotes. The first is the watermark which 75% of people recognise and the second is the hologram at 49%.[10]

Holograms can be used on both paper and plastic banknotes. They are not just used on banknotes—they are commonly found on credit cards and driver's licenses as well. In a report prepared by the National Research Council to assess the threat of counterfeiting to US currency, the effectiveness of holograms is discussed:

Because a hologram possesses extremely fine structural features, it is essentially impossible to copy or duplicate reprographically with even the most sophisticated equipment; hence, it is an excellent deterrent for the casual counterfeiter. However, it is possible to replicate or simulate hologram security devices using more advanced technical methods.

A major shortcoming is the hologram's lack of durability under even normal usage when placed on a flexible substrate. The image rather quickly becomes unrecognizable, making the job of a counterfeiter much easier, since a "worn" hologram is easy to simulate. The durability of holograms suffers from the wrinkling of the metallic film. If these devices were not metallized but were instead placed in a "window" on the banknote for viewing in transmission, their durability might be significantly increased. (Excerpt reproduced from ref. 11 with permission from National Academies Press, Copyright 1993.)

A hologram, like all optically variable devices, changes its appearance when viewed from a different angle. You can think of a hologram as a three-dimensional photograph that physically exists in only two dimensions. The word hologram is based on the Greek word *holos* for whole, and *gramma*, which refers to something drawn or written, such as an inscription. A regular photo provides a record of how light interacted with an object—it shows which wavelengths were absorbed and which ones were reflected; most importantly, it reveals information about the intensity of each particular wavelength of light that was reflected—if red waves were especially intense, your picture will be mostly red. The intensity of a wave is determined by its amplitude, or height of the wave.

A photo is really just a recording of the amplitude of the waves that were reflected from an object. A hologram, however, records more information than just amplitude. It also tells you whether or not those waves were lined up with one another. If two waves are in sync (in phase), their crests and their troughs will be in perfect alignment—they will constructively interfere. If they are not in step they destructively interfere. Therefore, a hologram records information not just about the amplitude but also about the interference patterns.

In order to record the precise way that waves of light interfere with one another, it is necessary to use laser light. Laser light is coherent—it only contains a small band of wavelengths of visible light. Coherent light tends to be monochromatic, as opposed to white light, which is polychromatic. You can always tell whether light is monochromatic or polychromatic by shining it through a prism or diffraction grating. White light will produce a rainbow of color, but laser light will produce light of only one color.

Even though the idea for the hologram was conceived in the 1940s, the creation of the hologram itself had to wait until the 1960s, when lasers were invented. To create a hologram, laser light is split in half using a half-silvered mirror, allowing for half of the light hitting it to be transmitted and the other half to be reflected. The reflected light will be aimed toward a mirror that will eventually enable it to meet up with the other half of the laser beam. Meanwhile, the other half of the beam travels toward another mirror, which bends the light toward the object you are trying to holograph—the light reflecting from this object eventually meets the other half of the laser beam. A hologram is

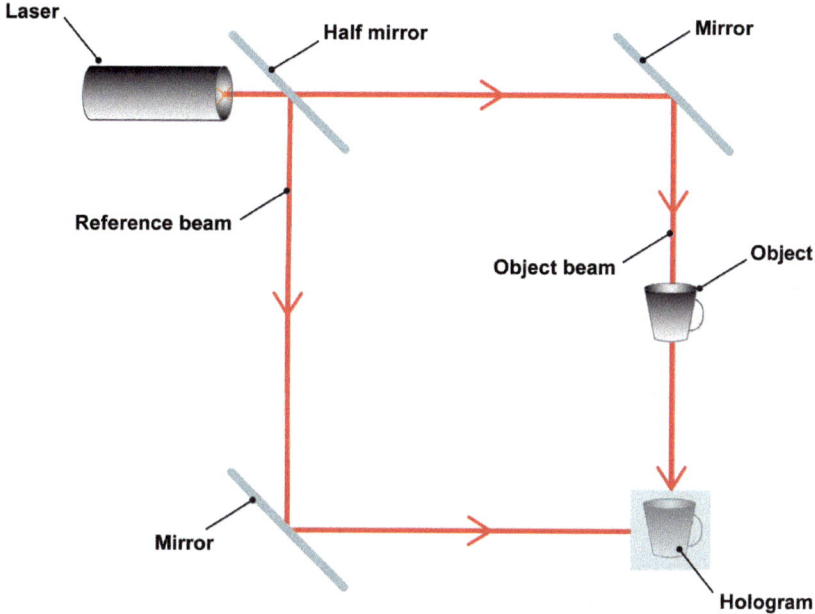

Figure 6.8 How a hologram is created.

formed where these two diverging beams meet up again. A special camera records the interference pattern formed by the light beams, creating a 3-D picture (Figure 6.8).

The holograms appearing on banknotes are typically of the rainbow variety, which can be viewed under white light, as opposed to other types that require a laser light to view. The holograms appearing on credit cards are also rainbow holograms. As light passes through the hologram it is reflected off the thin metallic film that provides the backing. These types of holograms are good examples of transmission holograms, as they allow the transmission of light—they can be viewed opposite the side from where light originated if the substrate is transparent, or in this case on the same side as light is reflected. Holograms are so intriguing because when you move your head or tilt the hologram in a different direction you can appear to see around the holographic object.

The method with which the hologram is applied to the banknote itself varies, depending on its type. Hologram threads are woven into the paper by the maker of the substrate, much like the polyester security threads of US bills. Stripes and patches are often hot stamped onto the substrate.

Hot stamping involves using a heated die to literally stamp a piece of metal foil containing the holographic image onto the paper or plastic and is often performed by the supplier of the substrate. The holographic images come in long rolls that contain a heat-activated adhesive, which along with pressure bond them to the paper permanently—they cannot be removed without damaging the banknote.

Laminating and rolling are other methods used to apply a foil hologram to banknotes. The holographic image can be embossed onto an aluminum layer, below which is an adhesive layer that is activated by heat. A protective lacquer layer lies atop the aluminum layer. After application the total thickness of the foil hologram will be from 3–5 µm. Laminating involves the application of light pressure and heat for a fairly long period of time, while rolling involves considerably more pressure, and heat, for a much shorter time period. The time of contact, referred to as the dwell time, varies inversely with pressure. For laminating, a typical dwell time of 420 ms is needed, at a pressure of 0.38 Mpa at 150 °C; rolling might entail a pressure as high as 50 Mpa at 235 °C for only 3 ms.[12] Whatever method is used, essentially the same process is used each time a hologram is affixed to a banknote—an adhesive is activated by heat and made to stick by pressure. Before the introduction of OVDs adhesives were not needed in banknotes—besides paper and ink, the only other element you might encounter would be the security thread, which was typically woven through the paper, not affixed to the surface.

A patent application gives the details of one type of adhesive used in a hot stamping foil:

> A preferred adhesive material, known as a "hot-tackifying adhesive," is described... The adhesive material is non-tacky or poorly tacky at 20 °C, but becomes pressure-sensitive and aggressively tacky when heated. Good bonds are immediately formed at a tackifying temperature without any need for crosslinking or other chemical reactions. The adhesive material comprises an acrylic polymer or mixture of acrylic polymers of at least one alkyl acrylate and/or methacrylate ester monomer.[13,14]

The layer of adhesive is thicker than that of the hologram foil itself. This thick layer is necessary so it can work its way down into the substrate, which is especially important on the rougher surfaces typically encountered on banknote papers—the rougher the surface the harder it is to get something to adhere as there are fewer surfaces in contact. It is easier to get glue to stick to a plastic banknote than a paper one, as plastic is much smoother.

6.13 MOIRÉ MAGNIFICATION

The US $100 bill is among the most counterfeited banknotes in the world (Figure 6.9). In 2013, a new feature was added in the hope of making it virtually counterfeit proof. A large 35-μm thick bluish-purple security strip runs through the width of the bill, just to the right of Ben Franklin's head. If you place it on a hard surface and press down firmly enough with your fingernail you can feel resistance when you come up to the edge of the strip, but otherwise it has a smooth plastic feel. There are little images contained within the strip, alternating between bells and the numeral "100". When the bill is held by the ends and rotated laterally, the images move from side-to-side, perpendicular to the axis of rotation. If the two ends of the bill are moved up and down, the images move along the length of the strip, again perpendicular to the axis of rotation. Not only do they move, but

Figure 6.9 One hundred (US) dollar bill.
© Shutterstock.

any movement of the images, either along the width or length of the strip, causes them to change back and forth from bells to the numeral "100". You can also see the same effects by keeping the bill constant and moving your head either up and down or back and forth.

According to the Bureau of Engraving and Printing, the new $100 banknote is the most complex US bill ever made. The strip is woven into the bill, not printed or glued on top of it—you can see white areas on the strip where it has been woven into the bill, much like the fluorescent security threads. However, if you look at the back of the bill you cannot see any portion of the strip. At first glance the strip looks like it contains a hologram, but it actually contains a multitude of little microlenses, roughly a million of them, each one so tiny that a dozen of them would fit on the tip of a human hair.[15]

A lenticular array, as a group of microlenses is called, is not as uncommon as you might think. Certain types of eyeglasses, such as bifocals, employ lenticular arrays. If you see a book cover or label that looks like it is in 3-D, or otherwise exhibiting some intriguing motion effects, it may be utilizing a lenticular display to create the illusion of shape-shifting or movement. These effects are only possible because the little lenses that make up the display are placed over the graphics in such a way that changing the angle at which you look through them changes the image you see.

The essential principle behind these features is moiré magnification, in which moiré interference between the microlens array and the underlying images produces the illusion of motion. Moiré interference patterns are easily observed whenever you have two gratings of closely spaced parallel lines superimposed on one another. The movement of one of the gratings ever so slightly produces a unique interference pattern of alternating dark and light lines that cannot be seen with either of the two gratings by themselves. The array of lenses is spaced over these underlying images so that when moved back and forth the illusion of motion is created.

The moiré effect is caused by parallax, the effect created when an object viewed from a different angle appears to change its position. The best way to demonstrate parallax is to hold up your index finger out in front of you at arm's length. If you close one eye and look at the finger it appears to be in one position, but if you close the other eye and look at the same finger it appears to have changed positions. The finger itself has not moved, of course, but it appears to because the position of the eye used to view it has changed. If you alternately close and open each eye rapidly as you look at your outstretched finger, it appears to be moving back and forth. In the same way, when you tilt the $100 bill and appear to see motion you are exploiting the effects of parallax.

The strip itself is composed of multiple layers of synthetic polymers. The tiny lenses are situated on the surface, and below these is a transparent layer that acts as a spacer between the lenses and the bottom layer, upon which is printed the images. As the middle layer is acting as an optical spacer, the thickness of this layer essentially represents the focal length of the lens. A polymer layer on top of the lenses seals them in, offering protection. Pretty much any type of transparent material can be used to make a lens—even a water droplet makes for a decent lens; the greater its refractive index the more it will bend light and thus the shorter its focal length can be.[16]

The lenses magnify the printed images below them hundreds of times. This tiny printing is much smaller than the other microprinting on the banknote, hence the need for the transparent optical spacer layer; any type of magnifying device—from a telescope to a microscope—requires a distance between the lens and the illuminated object so as to provide enough room for the refracted rays to converge and form an image. Each microlens is a miniature convex lens, which refracts light as it passes though, enabling the rays to eventually form the image you observe.

This technology first appeared in banknotes on the 2006 Swedish 1000 SEK note, and has since been steadily gaining in popularity—it can now be seen on various denominations around the world. The most popular application of moiré magnification is Crane Currency's Motion® technology, which is utilized in the US $100 note. Crane Currency envisions a bright future for their Motion® products. Their technology has the ability to create images that appear to be floating within a banknote, while others appear to jump out of its surface.[17]

6.14   TACTILE FEATURES

Although a lot of optically variable devices give the illusion of three-dimensionality, many banknotes today are actually becoming three-dimensional. If one chooses to be pedantic it could be argued that every banknote is three-dimensional since the ink itself has some height to it, especially the raised lettering produced by intaglio printing. However, today's banknotes go way beyond these relatively minor variations in topography, with the deliberate incorporation of features that are designed to be felt. These tactile features are becoming increasingly common on banknotes, running the gamut from raised lettering to raised dots.

On the polymer £10 banknote you can feel little groupings of raised dots in the upper left-hand corner; other denominations have a slightly different tactile pattern. Euro banknotes have a textured pattern along one edge, making them especially friendly to the visually impaired. Many countries are incorporating various combinations of raised dashes and dots on their banknotes, and even though they may resemble Braille, do not actually spell out any words or numbers. These tactile features are usually only incorporated on one side, so as to make for easier differentiation between the face and back of a bill.

Intaglio printing has always been able to produce raised lettering—the more deeply engraved the plate, the higher the lettering it can produce. A study by the De Nederlandsche Bank on banknote design for the visually impaired reveals the technology that has allowed for the incorporation of the more recent tactile features:

> New intaglio plate making techniques based on drilling or laser engraving became available around the year 2000. Both are using digital techniques instead of chemical etching. One of the major improvements to the traditional etching is the possibility of a higher relief and more sharp lines, especially relevant for tactile patterns. (Excerpt reproduced from ref. 18 with permission.)

Whereas traditional chemical etching of intaglio plates could only produce relief as deep as 50 μm, the new digital engraving systems are capable of producing relief as deep as 150 μm, which doesn't sound like much until you consider that the human fingertip can detect thicknesses as small as 13 nm![19]

6.15 MICROPRINTING

One of the more common security features on banknotes is right in front of your nose, but you won't see it unless you look very carefully, or better yet, use a powerful magnifying glass. Under magnification you are almost certain to see printing you could not see with the naked eye—what looks like a line or some other type of design just might be a row of tiny letters. You can usually find the denomination written in miniscule characters some-where on the bill, often in several places. Virtually every bank-note issued today will contain some type of microprinting (Figure 6.10).

On the redesigned polymer Australian $10 note banknote is-sued in 2017 you can see the portraits of two poets, that of AB 'Banjo' Paterson on the face and Dame Mary Gilbert on the back. What you won't see without a magnifying glass are excerpts from their poems in microprint. It is very difficult for counterfeiters to reproduce print that small. One of the easiest ways to detect a counterfeit is to magnify an area of the banknote that you know contains microprint; on all but the best counterfeits this writing will usually be blurred.

Figure 6.10 Microprinting close-up on a 50 euro note.
© Shutterstock.

Miniature writing has a long and rich history—tiny inscriptions recorded on Sumerian cuneiform tablets date back to the 3rd millennium BC. Hebrew scribes engaged in micrographia so they could add pictures to manuscripts. They created beautiful illustrations composed entirely of words, thus bypassing prohibitions against illustrating scripture.[20] In 1935, one man found the patience to write the entire Declaration of Independence, containing a total of 7576 letters, on a grain of rice![21] There is virtually no limit as to how small writing can be—in 1989 IBM amazed the world by using 35 xenon atoms to spell out their company's logo, visible only with a scanning tunneling microscope. Despite their endless promotion of this feat, complete with pictures, no one really "saw" the atoms.

We will never be able to see individual atoms because they are simply too small to reflect visible light. It would be akin to throwing a bowling ball at a speck of dust in the air and expecting it to be bounced back. The scanning tunneling microscope that produces images of atoms does not actually "see" them, but rather uses an electrically charged fine metal tip (just one or two atoms wide) to scan the surface of the sample. Fluctuations in the current are used to trace the resulting topography, revealing the presence of individual atoms. On the campus of Penn State University is a plaque in front of Osmund Laboratory commemorating the location in which atoms were first "seen" (Figure 6.11). No one saw those atoms either.

The unaided human eye can typically see objects as small as 0.1 mm, depending on your visual acuity. We can see that the microprinting is there—the problem lies with the eye's resolution, or ability to see the space between the tiny letters. If you are looking at two dots and they look like a single blob, your eye's resolution is not as great as that of someone that can clearly delineate the two dots. If you draw two lines very close to one another and move them farther away from your eyes they will eventually appear to merge into one line; at 15 cm away a person with good eyesight can generally make out two distinct lines that are just 0.04 mm apart; if your eyes are really good you might be able to see the space between two lines that are even 0.03 mm apart, but if the lines get much closer chances are they will appear as one.

Figure 6.11   First atom "seen" blue plaque at Penn State University.
Image courtesy of the Pennsylvania State University.

In their study of US currency, the National Research Council reveals the following details about microprinting:

On all but the $1 and $2 notes, the portrait is large enough to accommodate microprinting and fine-line details. Microprinting is used to print 0.2 mm tall letters with a line width of 0.05 mm. The production line width is approximately 0.1 mm, with a spacing of 0.1 mm. Microprinting and fine-line printing are used because they are difficult to reproduce with low-resolution electronic devices and can be viewed by the sharp-sighted or with a simple, low-power magnifier. (Excerpt from ref. 22 with permission from National Academies Press, Copyright 2006.)

To get an idea as to the tininess of this microprinting, set the font size on your word processing program to the smallest allowable size and then type a sentence. On the version of Microsoft Word I currently use, the smallest font size is 1-point, which equates to a height of 1/72 of an inch (0.353 mm). The microprinting on banknotes is even smaller than this. If you want to test the resolution of your printer, print out a portion of text in size-1 font—when I did this on my low-cost black and white laser printer, I couldn't make out any words; under magnification the letters were indiscriminate little blobs of ink.

Letters printed in size-2 font, though, were clearly distinguishable.

6.16   INKJET PRINTING

In the old days, if you wanted to print money you had to engrave your own plates and set up a printing press—it took a certain degree of expertise to be a respectable counterfeiter. However, with the advent of digital printing the capability was suddenly there for even an amateur to produce passable counterfeits. It is no accident that the proliferation of non-printable features on banknotes—such as holograms and the like—have coincided with advances in digital printer technology.

Digital printing involves two main types of technology—inkjet and laser printing. We will briefly examine each in an attempt to ascertain which might be best suited for printing money. Inkjet printers squirt out tiny droplets of ink as the paper moves past the print head. A typical inkjet printer will contain hundreds, or even thousands of tiny nozzles, all of which squirt out ink simultaneously (or in rapid succession). Each nozzle has an incredibly small opening—about 70 μm in diameter, making the early models especially susceptible to clogging. As a result, dye-based (water soluble) inks were used almost exclusively—as the colorants were dissolved the particle size would be much smaller than those found in pigment-based inks, and thus much less likely to clog up the nozzles. However, there are major drawbacks to printing with water-soluble inks—the printouts take a long time to dry, making them susceptible to smudging, and once they do dry, moisture can cause the ink to run. The ink also tends to spread out more, making printed characters thinner and shorter-lived.

As dye-based inks rapidly absorb into the paper, they can cause bleeding, a definite no-no if trying to print tiny letters. Pigment-based inks, on other hand, contain a volatile solvent that evaporates much more quickly. They are also more durable and less susceptible to fading, being especially resistant to UV radiation.

Today's higher resolution inkjet printers use pigment-based inks. As they don't absorb as deeply into the paper, they bleed less. Bleeding can also be controlled by using the right type of paper. Inkjet paper will tend to have a gloss, which will cause the ink to adsorb onto the surface rather than migrate too deeply into it. Using the right paper can double your resolution—whereas you might only achieve a resolution of 720 dpi (dots per inch) with normal copy paper, using inkjet paper can perhaps improve your resolution to 1440 dpi, depending on the type of printer used.

Smoother paper tends to be whiter, as there are fewer surfaces for light waves to bounce around in and become absorbed, as opposed to more porous, rougher paper that is a duller white as a result of absorbing more light. Duller types of paper tend to absorb ink like a sponge, making them especially unsuitable for inkjet printing, especially if you are trying to capture fine detail. Low absorption always means better resolution. If you examine even the tiniest microprinting on a banknote under the microscope, you will see virtually no bleeding onto the substrate. The bright white inkjet paper can be problematic, though, as it usually contains optical whiteners that make it much whiter than dull white, optically dead currency paper.

An inkjet printer can produce incredibly small dots—diameters as small as 50–60 μm are possible, each with a volume as tiny as 10 pL. As the microprinting on US currency is 0.2 mm tall (200 μm) a quality inkjet printer can theoretically do the job, provided the ink does not absorb too deeply into the paper and bleed, obliterating the fine detail.

If you have an inkjet printer at home, in all likelihood it is a bubble jet printer, which utilizes thermal energy to do its work. An electric current flowing through resistors raises the temperature to around 300 °C, heating the ink until it boils. As bubbles of ink vapor are produced, the ink is pushed through the nozzle. As the bubbles pass, a vacuum is created in the chamber directly above the nozzle—as nature abhors a vacuum ink rushes in to fill the void, ensuring its continuous flow. The spherical ink droplet flattens out as it hits the paper, spreading out as it is absorbed, forming a dot. A typical bubble jet printer can produce upwards of 300 million drops per second. Each bubble forms a single dot.

Another type of inkjet printer is of the piezoelectric variety, in which an applied voltage causes a tiny plate of piezoelectric material to contract, forcing ink out of the nozzle. This type of printhead is purely mechanical in operation, as no heating is involved. The piezoelectric effect occurs when mechanical stresses applied to certain crystals generates an electric charge. The igniter on a typical gas grill utilizes the piezoelectric effect—when you depress the button to start your grill a little spring-loaded hammer strikes a quartz crystal and generates a spark, igniting the gas. In inkjet printers, it is actually the inverse piezoelectric effect that is employed, whereby electrical energy causes the crystal to contract—as it deforms it forces a droplet from the nozzle.

Both the thermal bubblejet and piezoelectric inkjet are drop-on-demand printers, dispensing discrete little microdrops one at a time. The other type of inkjet printer is the continuous inkjet, in which a high-pressure pump forces ink through the nozzles in a steady stream—the piezoelectric effect breaks up this stream into itty-bitty droplets. Each drop is then electrostatically charged as it passes through a charging tunnel, where it is then deflected by an opposing charge.

The continuous inkjet printers are among the oldest inkjets on the market—they are primarily used for industrial purposes, finding their niche in the printing of various labels and markings on consumer products, such as expiration dates and bar codes. Even though a continuous inkjet printer is often used for low resolution work (owing to its remarkable speed and efficiency)—its tiny droplet size makes it more than capable of producing extremely detailed images, making its workhorse reputation somewhat deceiving.

6.17   LASER PRINTING

If you compare a printed page or image that emerges from an inkjet printer to one that comes from a laser printer, at first glance they may look the same, but the process from which they arise could not be more different. Laser printers are based on the principles of xerography, which literally means "dry writing", from the Greek roots *xeros* for dry and *graphia* for writing. Photocopiers rely on this same process; when the photocopier first came out a copy was referred to as a xerox, an innovation so enormously successful that the company that developed it would change its name to the Xerox Corporation. Laser printers also make use of xerography and grew out of photocopiers; a laser printer is basically a modified photocopier.

The heart of a laser printer is the photoconductive drum. If you have ever replaced the toner cartridge on your laser printer then you may have noticed the bright green cylinder attached to it, which is the drum. There is a reason it comes so tightly packaged—exposing it to light will ruin it. A bad drum is just as likely to be the cause of faded copies as a lack of toner; however, this is typically only a problem on models in which the drum is not attached to the toner cartridge—when they come packaged together the drum will usually outlast the toner.

In the early days of laser printing the drum would come coated with selenium, a semiconductor; selenium has largely been supplanted by organic photoconductors (OPCs)—synthetic plastics that are applied in a 20–50 µm thick layer on the surface. Like all plastics, OPCs are good insulators capable of holding a charge. Laser printing is initiated when the drum passes by a charge roller, which imparts a positive charge to the drum. A positive charge always involves a deficiency of electrons, while a negative charge indicates an excess—when considering charges, it's helpful to remember that it's only the electrons that move—the positive protons stay put, firmly ensconced in the nucleus. The charge roller imparts a positive charge by removing electrons.

When you press "Print" a message is sent to a laser, which bombards the drum with millions of laser pulses per second as it goes back and forth across its surface, tracing a mirror image of whatever words or images are to be printed. It is here that the photoconductive coating comes into play, which conducts electricity when exposed to light. As the inside of the drum contains a metal core that is electrically grounded, a ready reservoir of electrons lies just below its surface. As the parts of the drum that are bombarded by these laser pulses are all of a sudden conductive, electrons are now free to move into these formerly electron-deficient areas, rendering negative those parts of the drum that were struck by the laser, creating a perfect image of what is to be printed.

The negative electrostatic image on the drum now passes by another roller, which contains toner particles that have been positively charged. These positively charged particles stick to the negatively charged areas of the drum like glue. Next, a piece of paper, which has received a strong negative charge by passing by yet another roller, is fed toward the drum. As the negatively charged paper passes over the drum, the positively charged toner particles jump toward it, in the precise image first laid down by the laser. Finally, the paper passes through the fuser unit, which consists of two heated rollers in excess of 200 °C, hot enough to melt the toner particles. The rollers also apply sufficient pressure to fuse the melted particles into the paper, forming a permanent image about 15 μm high. In this final step, the solid toner particles essentially become ink, rendering the finished product virtually indistinguishable from one printed by typical "wet" ink.

The bulk of the toner is finely-ground plastic, since it needs to hold a static charge and have a low melting point. Polyester makes an excellent toner base—it is cheap, easy to make and has a low melting point. Just like ink, the other primary component in toner is the pigment; which is often composed of finely divided carbon black. Black-and-white laser printer cartridges sometimes contain magnetite (Fe_3O_4). You can verify the type of toner pigment you have by seeing how it responds to a strong magnet. (Just be careful not to inhale the fine toner dust, which can be quite hazardous.) In order to increase the resolution of laser printers, the size of the toner particles has steadily decreased over the years. Initially, toner particles were around 12 μm in diameter, whereas today they are half that size, or even smaller.

Color laser printers are a bit more complex, but typically involve repeating the same process with different colored toners, each of which has its own separate drum. Cyan, magenta, yellow and black are each added to the paper separately, in a step-by-step process.

In 1995, less than 1% of counterfeit money was produced with a digital printer. According to the Secret Service, that number is closer to 70% today.[23] As for which type of printer is best for printing counterfeit currency, the continuous-stream inkjet printers are probably the most capable of producing the highest quality images. However, these models are prohibitively expensive, and are typically used only for large scale commercial processes. There are plenty of drop-on-demand models capable of producing stunning images at a much lower cost. Inkjet printers are generally considered to be the best printers for reproducing high quality photos, with some of the better models capable of achieving resolutions of over 5000 dpi, twice that of a high-end laser printer. However, when considering resolution, the dots per inch is just one factor. Inkjet printers tend to have trouble faithfully reproducing the fine intaglio-printed lines of currency, as their ink dispensing methods can make lines look a bit jagged. As currency paper is also somewhat rough, the bleeding from the ink used in inkjets can be a problem. However, most of the mom-and-pop counterfeiting operations you read about in your local newspaper revolve around an inkjet printer, as they typically cost a lot less than a color laser printer. Even though high-quality laser printers can do a great job of capturing fine details, there is at least one compelling reason why they might be avoided by counterfeiters.

If you look closely at a color laser copy you will notice faint yellow groupings of dots. They are hard to see, but if you use a blue light (or a black light) they will be more visible. The dots appear every few centimeters throughout any color laser printout, and represent a coded message containing the date, time and serial number of the printer used. Nearly every type of colored laser printer utilizes this tracking system.[24] One study revealed that these tracking dots were produced by 141 different printers from 106 different models produced by 18 manufacturers, representing the majority of the world's printers.[25]

There has been no shortage of public angst over these dots, with privacy advocates and concerned citizens crying foul. In an effort to get answers, the following question was publicly addressed to the European Parliament on November 20, 2007:

> Press reports have indicated for several years that a number of photocopiers and colour laser printers that are sold to consumers in the European Union contain forensic tracking mechanisms. The existence of these mechanisms has been disclosed by some manufacturers, but not by others. Press reports indicate that these devices invisibly print patterns of small yellow dots on all output documents, and that the patterns of these dots could aid in determining the origin of any such document. Some manufacturers have stated that these measures were implemented to deter

counterfeiting. Manufacturers have not publicly described how the tracking codes work or what information is coded. Recent research by civil society indicates that some printers and photocopiers are coding their serial numbers and the date and time of printing into each page, and that this information could be read by private individuals, as well as by public authorities. Some consumers have viewed the presence of tracking codes as an invasion of privacy and have unsuccessfully asked manufacturers to disable this function.

Is the Commission aware of any legal framework or obligations in Community law or national legislation relative to the use of these tracking mechanisms? Does the Commission believe that the current practices of manufacturers in this regard, including their disclosures to consumers, are consistent with relevant Community law on data protection and consumer protection?

The Parliament's answer is as follows:

The Commission is not aware of any specific laws either at national or at Community level governing tracking mechanisms in colour laser printers and photocopiers. In the cases outlined in the Honourable Member's question, the information based on tracking printed or copied material does not necessarily include data relating to identified or identifiable individuals, *i.e.* personal data.

To the extent that individuals may be identified through material printed or copied using certain equipment, such processing may give rise to the violation of fundamental human rights, namely the right to privacy and private life. It also might violate the right to protection of personal data.

The protection of privacy is ensured by Article 8 of the Convention of Human Rights and Fundamental Freedoms. The Charter of Fundamental Rights of the European Union, in Article 7, provides for the protection of private and family life, home and communication, and in Article 8, for the protection of personal data.[26]

The choice to make these dots yellow was no accident. Yellow has very little contrast with white, making it difficult to see.

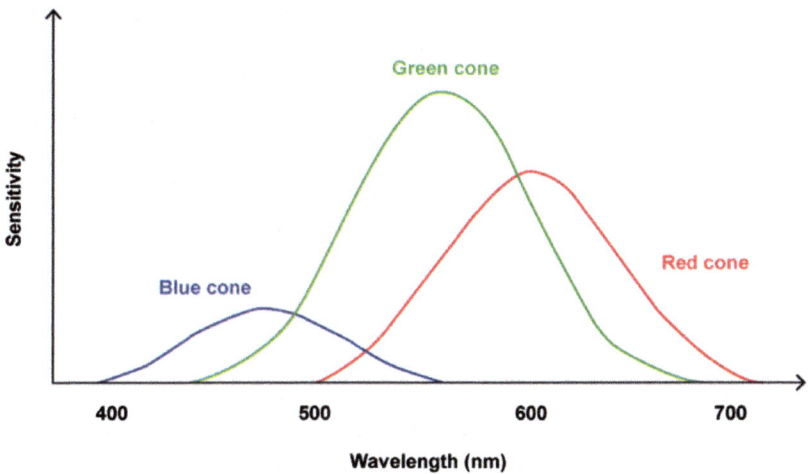

Figure 6.12 Sensitivity of our eyes to various forms of light.

Of the three types of cone cells in the retina, 64% of them are sensitive to longer wavelengths, which we perceive as red. 32% of them are sensitive to medium wavelengths, which we perceive as green light (Figure 6.12). Only 2% of cone cells are sensitive to blue light. When white light enters our eyes, all of the cones are stimulated at once. As yellow is a secondary light color that is made by combining red and green, when yellow light strikes our eyes, both the red and green cones are stimulated, representing 98% of the cone cells. Seeing yellow light is almost the same thing as seeing white light, at least from the point of view of our brain, which processes the signals it receives *via* the optic nerve.

6.18 CAN YOU PHOTOCOPY MONEY?

As photocopiers and laser printers rely on the same basic principles of operation, advances in one have paralleled advances in the other. As the ability of printers to print lifelike images has increased, so has that of photocopiers. Today's high-end copiers can perform a plethora of tasks, making it possible to for even the laziest of counterfeiters, armed with the right paper, to make startlingly realistic reproductions of money with just the push of a button. About the only thing today's copiers won't do is cut a sheet of money into individual bills.

Those involved in producing the world's banknotes are of course well aware of the capabilities of modern photocopiers, and are continually devising ways to make it difficult to copy their work. One particular feature, which was used on South African currency in 1992, is known as STRAP, an acronym for Système de Transfert Réfléchissant Anti-Photocopie (Figure 6.13). It consists of a series of metallic foil strips that can be incorporated into a banknote—when photocopied black stripes appear on the copies.[27]

Any shiny item will appear black when copied. If you're not convinced, try photocopying a mirror. Once you've tried it, then try a piece of piece of aluminum foil. Both will appear black. When the bright light from the photocopier scans an image, white or shiny areas reflect light and darker areas absorb light. The reflected light exposes areas of the drum; the exposed areas will attract toner particles, which will eventually be transferred to a piece of paper. All the light reflected from a mirror will result in a large area of the drum being exposed, which will translate into a big black area on a piece of paper. The use of foil has become ubiquitous on banknotes, and not only serves as backings for holograms and other optically variable devices but can be a valuable security feature in its own right, as its reflective surfaces make them not only highly visible but also difficult to photocopy.

Figure 6.13   South African Reserve Bank note.
               © Shutterstock.

Reflective strips are only one device used to preventing photocopying. If you have ever tried to make a photocopy of a sheet from a doctor's prescription pad, the copy will come out with dire warnings, like "DO NOT COPY" or "VOID," splashed across its surface. This technique, referred to as a void pantograph, is commonly used in the security industry. The term *pantograph* refers to an old-fashioned device once used to make copies or enlargements of maps or other documents. You would put your pen in one side of it and another pen on the other side, and as you wrote a mirror image would be produced, the size of which could be adjusted. It resembles a folding barrier you might erect over a storefront, or maybe like the side of a foldable laundry rack.

The word *VOID* can be made to show up on a copy, but not in the original, by creating a pattern of dots of varying darkness in the background of a document. This pattern will be extremely difficult to see with the naked eye—it may not even be visible at all. But a copier or scanner will recognize the pattern and will assume this is what you meant to print, spitting out a copy that looks quite a bit different than the original. If a document has a fancy background, there is a good chance it is utilizing a void pantograph. Take a look at your official academic transcript—the ornate background is not there to convince you that you attended an august institution, but rather to prevent you from committing fraud. If you try to photocopy your transcript, a hidden message will no doubt appear on the copy.

Void pantograph patterns can be circumvented on a lot of copiers by simply changing the density settings, and some copiers might not even copy the pattern at all, producing copies truer to what the eyes see. As with banknotes, most printers of security paper use a variety of techniques to prevent photocopying, so they won't just rely on a void pantograph or some other singular measure; they might include watermarks, microprinting, color-changing inks or some other device.

To my knowledge, a void pantograph has never been used on banknotes (though it is commonly used on checks), as far more advanced techniques have been developed to prevent you from running down to the local copy shop whenever you find yourself running low on cash. I had always heard you could not copy banknotes, and research from a plethora of sources confirmed this sentiment. But like a good scientist, I had to see for myself.

I was already in the local library when I decided to see what would happen if tried to photocopy money. Making sure no one was watching, I started with the face of a US $20 bill, and 50¢ later I had a beautifully rendered full color copy on my hands—the detail, and especially the color, was incredibly striking—I could easily have passed it off for the real thing if it were printed on the right kind of paper. But when I tried to reproduce the back, the copy came out pure black. It seemed odd that that I could copy the face but not the back, but I guess a one-sided counterfeit is pretty worthless. Yet when I tried copying other bills—I had a $5 and a $100 bill on me—neither side could be copied; they all came out pure black. I couldn't even print them in black and white. The only bill I could print in color—both face and back—was the lowly $1 bill.

Explanations abound as to why you can't copy money. One theory that was bandied about at the turn of the century was that every banknote contained a geometric pattern of five colored dots that software within the copier was programmed to read, and if detected, would block reproduction of the image. This pattern was dubbed the EURion constellation, as it looked like the actual constellation Orion (Figure 6.14).

Figure 6.14   Orion constellation.
        © Shutterstock.

Reproduction of the image, in color but not black and white, has been shown to prevent copying on multiple occasions.[28] Other research has revealed that the dots must be 1 mm in diameter and be precisely spaced in order to prevent photocopying. The exact orientation of the pattern itself does not seem to matter.[29]

The idea that the same pattern appears on all the world's banknotes, and that this pattern can prevent them from being copied, has all the makings of a conspiracy theory, and a weak one at that. However, no conspiracy theory is complete without either unequivocal denials or withering silence by the governing authorities, which this one has. Today's banknotes contain so much noise that it wouldn't be that hard to find virtually any pattern of dots you might be looking for, in much the same way that you can look up at the stars and make all sorts of patterns out of seeming randomness.

Assuming you were able to find such a pattern on a banknote, it would be difficult to prove that it was this pattern that prevented it from being copied, especially without the cooperation of those who produced it. Correlation does not imply causation. Yet, there must be some reason I couldn't make a copy of that $100 bill in my pocket. What that reason is, however, we may never know.

There are numerous patents on file detailing pattern detection methods that can be incorporated into a banknote.[30,31] It is no secret that numerous level three security features exist within our banknotes that can only be detected by specialized scanners. In their report on counterfeit deterrence, the US National Research Council provides some insight as to what these scanners are searching for:

> The features of US banknotes most used in current machine readers are the optical spectrum and image, magnetic inks, ultraviolet fluorescence, ultraviolet spectrum, and infrared ink pattern. Low-end readers may sense only a single feature, usually the infrared ink pattern; high-end readers may use 10 or more measurements to authenticate each note.
>
> Typically, manufacturers of machine readers report that low-quality counterfeits are identified by a low optical

image quality, lack of magnetic and/or infrared ink, or incorrect paper fluorescence. High-quality counterfeits may require detailed magnetic signature sensing or ultraviolet spectrum sensing to be detected. Ninety percent of suspect notes are caught because they do not have an authentic magnetic pattern signature.

Most overt counterfeit-deterrent features are not used by machine readers because they are difficult to sense, locate, or verify. Features not used for machine authentication include color-shifting ink, cotton fibers, watermarks, security strips, and microprinting.

In a high-speed counter, the signal strength from each note must be high enough to be sensed in 0.04 seconds. Several features can be useful for reading in this short time, including the magnetic ink signal, the pattern in infrared ink, and the ultraviolet spectrum and fluorescence of the paper. High color contrast, as in previous, all-intaglio note series, provides a strong optical signal; however, artistically smooth shadings and the addition of multiple colors and features cause the optical contrast to decrease. In addition, for the purposes of machine readers, the signal from a feature must be reliable. Overall color, which changes with use, is an example of an unreliable signal.

Certain features on current notes are easily read in either feed direction. These include the ultraviolet spectrum and fluorescence of the paper and the patterns in magnetic and optical ink in the printed image. An example of a feature that is not readable in both directions is the infrared ink pattern, which is a set of stripes parallel to the short edge of the note. (Excerpt from ref. 32 with permission from National Academies Press, Copyright 2006.)

Even though color printers and photocopiers have become more advanced, there are still any number of features on banknotes that do not copy well—you can't photocopy a watermark or a hologram or color-shifting ink. Despite all of these drawbacks, a decent high-resolution digital printer or copier can still capture enough detail to make a copy passable, and besides, counterfeiters are counting on people being ignorant of the details, or more likely, just not paying attention.

6.19 ANALYTICAL TECHNIQUES

Just as the central banks have methods of counterfeit detection that are inscrutable to the masses, so too do chemists—they have at their disposal a whole arsenal of equipment and techniques that can expose even the best attempt at forgery. A chemist can tell you the exact composition of pretty much anything by employing any one of a number of methods; unfortunately, some of these methods can destroy the very thing you are testing—it is little consolation to find out that your $100 bill is genuine if you had to reduce it to ashes to do so. Fortunately, there are several non-destructive techniques that have shown great promise in detecting counterfeit currency.

Desorption electrospray ionization (DESI) involves a deceptively simple-sounding procedure that entails spraying the surface of a substance to be tested with very fine electrically charged droplets of a suitable solvent, an electromist if you will, and then collecting the residue that bounces off. These offspring droplets take with them particles of the substrate, which have now been desorbed (the opposite of absorbed). This process is somewhat akin to taking a pressure washer to a surface and blasting away— if you were to analyze the runoff you would surely find little bits of whatever it was you were blasting at. These desorbed droplets make their way into a mass spectrometer, a wonderful device that ionizes whatever you put into it and then accelerates it rapidly through a magnetic field. As a magnetic field will deflect lighter particles more than heavier ones, a characteristic spectrum based on mass can be produced. If the mass spectrum of a suspect sample is significantly different from that of an authentic sample, then you are probably looking at a counterfeit.

One study focused on using the DESI technique to see if differences in Brazilian reals and known counterfeits could be distinguished from one another. The spectra of each was noticeably different, as shown in Figure 6.15.[33]

The differences in spectra between genuine and counterfeit currency were primarily caused by differences in the inks. Researchers were even able to isolate the specific brand of toner used in a laser printer to print a counterfeit. Similar results were also obtained when American and euro banknotes were analyzed alongside counterfeited bills. Other analytical techniques

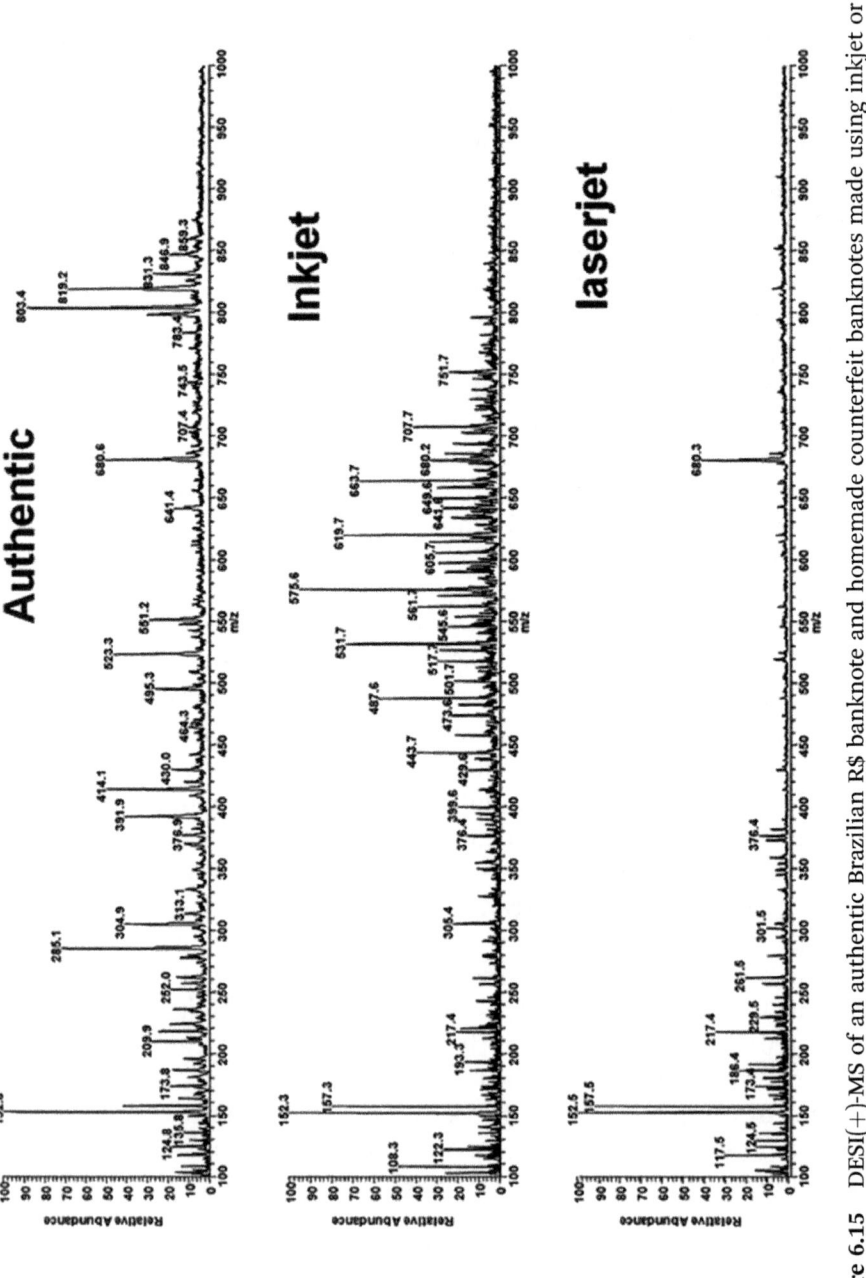

Figure 6.15 DESI(+)-MS of an authentic Brazilian R$ banknote and homemade counterfeit banknotes made using inkjet or laser jet printers.
Reproduced from ref. 33 with permission from the Royal Society of Chemistry.

have also been successfully used to detect counterfeit currency, such as IR spectroscopy, gas chromatography-mass spectrometry and laser desorption mass spectrometry.

6.20   BLEACHED BILLS

Even the most dim-witted counterfeiter knows that it's all about the paper. If nothing else, the wide use of the iodine starch marking pen has convinced counterfeiters that they had better not use regular paper—starch-free is the way to go. Even though you can buy starch-free paper, it's not the same as currency paper. Therefore, unless you're passing a fake bill to someone with gloves on, the surest way to arouse suspicion is to present a bill that just doesn't feel right. The best way to simulate the look and feel of real currency paper is to use the genuine thing.

Raised notes, as they are called, are nothing new. One of the earliest surviving notes of the American colonies, dating from 1690, is a bill that was raised from 5 to 20 shillings. In 1690 paper money was still in its infancy, and as a whole host of institutions could print money of any number of denominations, raising notes was commonplace—it would be hard to pick out a forgery as there were so many different varieties of money in circulation. Although it is necessary to continually change certain features of banknotes so as to keep ahead of the counterfeiters, it is still important that each banknote retain its own individual identity—a $10 note should retain the look of a $10 note—too much change opens up doors for forgery, as a populace not yet familiar with the look of each denomination can easily be duped by someone making minor changes to a bill to increase its value. Whenever I travel outside of the US I have to remember that I am an easy mark for a counterfeiter—I probably will not have had enough experience with the currency of my host country to make any kind of informed decision as to whether or not a particular banknote looks genuine.

Probably the easiest way to raise a note is to bleach the ink off a lower denomination bill and then print one of a higher denomination on top of it. In 2012 the US Secret Service nabbed a man in Florida, USA, for bleaching the ink from $5 bills and using the paper to print $100 bills. When they raided his home, they found several bottles of chemicals that were used to remove the ink, as well as several counterfeit bills. He was using a digital printer to make the images.[34] Also in 2012, a Virginia, USA, mom was arrested for making $15,000 worth of counterfeit money using bleached bills. She would remove the ink from genuine banknotes by soaking them in a powerful degreasing agent, putting them in the microwave and then scrubbing off the ink with a toothbrush. Using scanned images, she would then use the bleached paper to print either $50 or $100 bills onto the $5 blank bills.[35]

The most popular bleached US banknote is the $5 bill as it has a fluorescent security thread and watermarks, and even though they don't correspond to those of higher denominations, the presence of these features might be enough to convince an unsuspecting person that a bill printed on this paper is genuine. They will also pass the iodine pen test.

As an academic exercise I once tried to see if I could remove the ink from a dollar bill. I first soaked the bill in household bleach, which did nothing. Next, I soaked it in an engine degreaser, and again nothing happened. It was not as easy as I had been led to believe, and lacking motivation, I abandoned my efforts.

One popular method for removing ink is to use a combination of acetone and bleach. I would most definitely not recommend this procedure, as these substances react to produce chloroform ($CHCl_3$), a volatile liquid that is a powerful anesthetic (in large enough concentrations). Other methods call for using degreasers—if they can remove the nonpolar grease from an engine, then they just might work on removing nonpolar ink from a banknote. Who knows—if I had applied a little more elbow grease to my little experiment with the engine degreaser, I just might have been successful.

Probably the best way to prevent people from raising notes is to simply make each denomination a different size, just like with coins. Making a $100 bill the exact same size as a $1 bill is asking for trouble. This is common practice in many other countries. The £5 banknote measures 12.5×6.5 cm, while the £10 note is 13.2×6.9 cm. The £20 note is 14.9×8.0 cm, while the £50 banknote is 15.6×8.5 cm. The same progression applies to the euro—the sizes get progressively larger as the denomination increases; the €5 note is among the smallest in the world. An added benefit is that the size differences help the visually impaired to distinguish between banknotes of different denominations.

When the euro was introduced in 2002, its differing sizes caused problems. The Dutch, for example, were used to the relatively small uniformly sized guilders with a height of 7.6 cm. The euros, especially those of larger denominations, stuck out of their rather small wallets, crumpling the bills, which not only looked unsightly but decreased their lifespan. Banks in the Netherlands subsequently reported more wear in the higher denomination euros than in the lower ones.[36]

6.21 EXPLODING DYE PACKS

If one is criminally inclined, there are easier ways to make a score than printing your own money—you can always rob a bank. Even if you successfully pull off a heist there's a good chance you won't get far with your stolen loot before encountering an unpleasant surprise. If you watch enough television, or work in a bank, you know all about the exploding dye packs that can permanently stain stolen currency. They are usually made out of a hollowed-out stack of $10 or $20-dollar bills, creating a cavity big enough to contain the exploding dye component. During a bank robbery the teller will slip one of these packs into the robber's loot, and when they pass through the door a radio transmitter sends a signal to a receiver within the dye pack to set off an explosive device 10 seconds later.

When the pack detonates, a bright red cloud of permanent dye is released, indelibly marking both the dye and the robber. Sometimes tear gas is released along with the dye. If you ever see a banknote with a little pink dye along the edges, it may very well have been marked with one of these exploding packs. The red dye is typically made of 1-(methylamino)anthraquinone ($C_{15}H_{11}NO_2$), also going by the aptly named Disperse Red 9. It is commonly used as a biological stain, as well as to color the smoke in smoke grenades. As it is a nonpolar substance, it is extremely difficult to remove.

Dye packs don't have to release a colored dye—sometimes the dye is clear. A whole line of products, marketed under the name of SmartWater, are designed to mark money (or criminals) with an invisible liquid that fluoresces a bright yellow under a blacklight. Shipments of cash or other valuables can be marked with this dye, and in the event of a robbery, the stolen goods can be identified if recovered. A number of ATMs in the UK are set up to release this dye all over anyone who tries to break into one, covering them and any stolen money with this highly fluorescent liquid. It is almost impossible to remove every trace of it, and just a speck can be enough to link a suspect to a crime. Smart-Water dyes are specially formulated for each customer (hence their name)—each batch contains unique markers that can be used to trace back a criminal or stolen money to a specific location.

6.22 COUNTERFEITERS AT WORK

Just like the counterfeiters of yore, professional counterfeiting rings today still employ heavy-duty printing technology, except now they have a lot more tools at their disposal, utilizing a combination of offset printing and high-end inkjet printers to achieve the best results. The best counterfeit bills, though, are produced by rogue nations and terrorist groups who can afford to purchase the same types of intaglio printing presses and other commercial equipment as legitimate currency producers, enabling them to reproduce even the latest security features. However, for the casual counterfeiter, a good inkjet printer is usually sufficient. The first step is to take a picture of the banknote, or if you can get past the software, scan an image and download it to a computer. It is not always possible to capture fine detail, so often the image will need to be touched up using a graphics program. If using a printing press, these images will then be burned onto plates.

Authorities that raided one counterfeiter's print shop confiscated enlarged negatives that were being manually altered. The best counterfeiters are often the best artists—some stunningly realistic banknote reproductions have been painted by hand.

Beginning in the 1990s a number of so-called "supernotes"—counterfeit US $100 bills—were showing up with alarming frequency; these were the best the authorities had ever seen. The US government tried to place the blame on North Korea, but as far as we know the origin of the bills is still unknown; the majority of these notes have not been circulated on US soil, however. Unlike most counterfeits, these bills were printed with an intaglio press, producing the distinctive raised printing that is such a distinguishing feature of genuine currency. Red and blue fibers were incorporated into the paper, just like the real thing. Even the watermarks and security strips were there. The level of detail in the printing was astounding. Not only did they use magnetic ink, but they were able to replicate color-shifting ink as well. Despite all this attention to detail, there were noticeable flaws that a trained eye could detect. Even the best counterfeits aren't perfect.

There are all sorts of ways to simulate the look and feel of genuine currency. Security threads can be sewn into the banknote by hand. Foil patches and strips that resemble holograms and the like can be glued on. Ultraviolet features are easy to forge—fluorescent highlighters and invisible UV pens can produce marks that glow brilliantly under a blacklight. Flour can be mixed with glue and water and rubbed over lettering to produce a glossy, raised effect. Automotive paint, which is often iridescent, can be used to replicate the color-changing ink.

I recently had the good fortune to stumble across a $20 counterfeit bill. I borrowed it from a friend of mine, who obtained it from his son who received it while delivering pizzas. My friend assured me that it was not against the law to possess a counterfeit bill as long as you don't try to pass it. It was exceptionally well rendered—it had the look and feel of genuine currency paper, with a mass of exactly 1.0 grams. Under magnification every bit of the microprinting was crisp and sharp. Holding it up to the light, I could clearly make out a distinct watermark. The only flaw I could see was that in the lower right corner, the color-shifting ink of the "20" did not shift—it looked the same regardless of the angle at which it was tilted. It was still a copper-green metallic color, though and would easily have escaped detection if you weren't looking for it.

Under the blacklight, though, the fakery became more evident. Sure enough, there was a fluorescent strip, but it was not embedded into the paper; it looked like it was made by marking along the top of the bill with a fluorescent marker. The biggest giveaway was the lack of microprinting on the strip, which is only visible under UV light on a genuine note—the counterfeiter didn't even try to replicate this feature. A blacklight also revealed some other interesting anomalies. On a genuine $20 bill, the only fluorescent aspect is the security strip—everything else is decidedly non-fluorescent. On the counterfeit, all the corners and borders glowed white under the blacklight, as well as a large swath to the right of the portrait. As I suspected this to be a bleached bill, my best hypothesis was that perhaps the bleaching agents caused areas of the bill to fluoresce that were not then subsequently covered with ink. To identify this counterfeit, I used both level one and two security features—the color-shifting ink was clearly absent, with the blacklight providing additional verification. If this bill was scanned at a bank, I'm sure it would have failed their tests as well.

After seeing this counterfeit and how closely it resembled the real thing, I have most likely had others in my possession without realizing it. I'm now on a mission to see if I can spot a counterfeit in my change. When I receive a banknote, I always hold it up to the light and study the details, breaking out the blacklight if I have any doubt as to its authenticity. If I ever do discover a real counterfeit, it is going to set up a genuine moral dilemma—if I turn it in to a bank or the police I lose the money, but if I spend it I'm guilty of a crime, since I have knowingly passed a counterfeit bill. Maybe ignorance is bliss—a little knowledge can be a dangerous thing.

REFERENCES

1. H. Potter, *The office and duty of a justice of the peace: And a guide to sheriffs, coroners, clerks, constables, and other civil officers: according to the laws of North-Carolina*, J. Gales Raleigh, 1816, p. 67.
2. P. Handler, Forgery and the end of the 'bloody code' in early nineteenth-century England, *Hist. J.*, 2005, **48**, 683–702.
3. J. L. Roberts, The veins of Pennsylvania: Benjamin Franklin's nature-print currency, *Grey Room*, 2018, **69**, 50–79.
4. D. Barrett, "Modest enquiry" and major innovation: Franklin's early American currency, *Visible Lang.*, 1995, **29**, 316–362.

5. T. H. Chia and M. J. Levene, Detection of counterfeit U.S. paper money using intrinsic fluorescence lifetime, *Opt. Express*, 2009, **17**(24), 22054–22061.

6. A. Bruna, G. Farinella, G. Guarnera and S. Battiato, Forgery detection and value identification of euro banknotes, *Sensors*, 2013, **13**(2), 2515–2529.

7. Sterling & Currency, Australia's 1988 bicentennial $10 note – an elaborate field test, https://www.sterlingcurrency.com.au/research/australias-1988-bicentennial-10-note---elaborate-field-test.

8. R. A. Lee, Diffraction grating and method of manufacture, *U. S. Pat.*, 5,428,479, 1995.

9. A. N. Becidyan, The chemistry and physics of special-effect pigments and colorants for inks and coatings, *Paint Coat. Ind.*, 2003, **19**, 6.

10. ITW Security Division, The hologram – still going strong!, An ITW Security Division White Paper, 2017, https://www.itwsecuritydivision.com/Portals/0/documents/ITW%20White%20Paper%20-%20The%20Hologram%20Still%20Going%20Strong.pdf?ver=2017-09-06-081840-913.

11. National Research Council (U.S.) and G. T. Sincerbox, *Counterfeit deterrent features for the next-generation currency design*, National Academy Press, Washington, DC, 1993, p. 69.

12. K. J. Schell, H. Klinker and R. L. van Renesse, A new method for testing the adherence between foil and substrate, Conference on Optical Security and Counterfeit Deterrence Techniques IV, San Jose, Ca., 2002, SPIE vol. 4677, pp. 169–174.

13. K. L. Hwang, Hot stamping foil, *U. S. Pat.*, US 6,514,586 B1, 2003.

14. R. J. Tollefson, J. G. Berg and P. D. Hinderaker, Sheet material adapted to provide long-lived stable adhesive-bonded electrical connections, *U. S. Pat.*, 4,569,877, 1986.

15. A. McCabe, Like magic: The tech that goes into making money harder to fake. NPR: All Tech Considered, 2017, https://www.npr.org/sections/alltechconsidered/2017/10/23/559092168/like-magic-the-tech-that-goes-into-making-money-harder-to-fake.

16. V. J. Cadarso, S. Chosson, K. Sidler, R. D. Hersch and J. Brugger, High-resolution 1D moirés as counterfeit security features, *Light: Sci. Appl.*, 2013, 2, e86.

17. Crane Currency, Reimaging banknote security – Motion Surface™. Banknote Technology Report, 2018, https://www.cranecurrency.com/wp-content/uploads/2018/06/BTR-SURFACE-4-18.pdf.

18. H. de Heij, Banknote design for the visually impaired. Occasional Studies, Vol. 7/No. 2. De Nederlandsche Bank NV, Amsterdam, 157, 2009, https://www.dnb.nl/binaries/Banknote%20design%20for%20the%20visually%20impaired_tcm46-224150.pdf.

19. L. Skedung, M. Arvidsson, J. Y. Chung, C. M. Stafford, B. Berglund and M. W. Rutland, Feeling small: Exploring the tactile perception limits, *Sci. Rep.*, 2013, **3**, 1.

20. J. Foer, A minor history of miniature writing, *Cabinet: A quarterly of art and culture*, **25**, 2007, http://www.cabinetmagazine.org/issues/25/foer.php.

21. L. F. Pfaller and J. Hjort, Bicentennial story #7-miniature writer. Southwestern North Dakota Digital Archive at the Dickinson Museum Center. Stark County Historical Society, 1975, https://dmc.omeka.net/items/show/396.

22. National Research Council (U.S.), *Is that real?: Identification and assessment of the counterfeiting threat for U.S. banknotes*, National Academies Press, Washington, D.C, 2006, p. 56.

23. A. Murakami-Fester, Counterfeit cash gets harder to spot: Some telltale signs, USA Today, 2016, https://www.usatoday.com/story/money/personalfinance/2016/10/23/counterfeit-money-spot-fake/92080738/#.

24. J. Tuohey, Government uses color laser printing technology to track documents, PC World, 2004, https://www.pcworld.com/article/118664/article.html.

25. R. Timo, E. Stephan, S. Dagmar and S. Thorsten, Proceedings of the 6th ACM Workshop/Information Hiding and Multimedia Security (IH&MMSec '18), Forensic analysis and anonymisation of printed documents, ACM, New York, NY, USA, 2018.

26. S. Hassi, European Parliament, Parliamentary questions, 2007, http://www.europarl.europa.eu/sides/getDoc.do?reference=E-2007-5724&language=EN.

27. R. L. van Renesseand VanRenesse Consulting, *Protection of high security documents: Developments in holography to secure*

the future market and serve the public, Holopack. Holoprint, Vienna, Austria, 2006, p. 10.

28. C. Baraniuk, The secret codes of British banknotes. BBC News, 2016, http://www.bbc.com/future/story/20150624-the-secret-codes-of-british-banknotes.

29. J. Nieves, I. Ruiz-Agundez, P. G. Bringas and 2010 21st International Conference on Database and Expert Systems Applications (DEXA), Recognizing banknote patterns for protecting economic transactions, 2010, pp. 247–249.

30. C. Baraniuk, The secret codes of British banknotes, BBC News, 2016, http://www.bbc.com/future/story/20150624-the-secret-codes-of-british-banknotes.

31. J. Nieves, I. Ruiz-Agundez, P. G. Bringas and 2010 21st International Conference on Database and Expert Systems Applications (DEXA), Recognizing banknote patterns for protecting economic transactions, 2010, pp. 247–249.

32. National Research Council (U.S.), *Is that real?: Identification and Assessment of the Counterfeiting Threat for U.S. Banknotes*, National Academies Press, Washington, D.C, vol. 12, 2006, p. 13.

33. L. S. Eberlin, R. Haddad, N. R. C. Sarabia, R. G. Cosso, D. R. J. Maia, A. O. Maldaner, J. J. Zacca and M. N. Eberlin, Instantaneous chemical profiles of banknotes by ambient mass spectrometry, *Analyst*, 2010, **135**(10), 2533–2539.

34. A. Pavuk, Counterfeiters turn bleached small bills into $100s. Orlando Sentinel, 2012, http://articles.orlandosentinel.com/2012-09-10/news/os-bleached-money-arrests-20120907_1_counterfeit-bills-counterfeit-money-genuine-money.

35. F. Green, Richmond-area counterfeiting scheme turned $5 bills into $50s, Richmond Times-Dispatch, 2013, https://www.dailypress.com/news/dp-xpm-20131209-2013-12-09-dp-wire-counterfeiting-scheme-1209-20131209-story.html.

36. H. de Heij, Banknote dimensions and orientation: User requirements. Presentation to the BPC/General Meeting on behalf of the BPC/Paper Committee, De Nederlandsche Bank NV, Krakow, Sept. 11–14, 2006.

The Future of Money

"There will be a time – I don't know when,
I can't give you a date – when physical money
is just going to cease to exist."

Robert Reich, former US Secretary of Labor

I recently ate lunch in the cafeteria of a local college. When I handed the cashier a $20 bill she hesitated, as if not quite knowing what to do with it. I asked if I was the first person to pay in cash that day and she assured me that I was not—one other person had also paid with cash.

It is not that uncommon to encounter establishments where cash is no longer accepted, an unthinkable proposition just a few short years ago. Sweden is nearly cashless, with 99% of purchases made without physical money—you can't even get cash from most banks. The easiest way to pay for things in Sweden is to use the mobile phone app Swish, which connects your bank account to your phone number. To pay someone you just type in their phone number and your PIN number, and in seconds money is transferred from your account into theirs. It can be used to pay businesses or individuals. Others are following Sweden's lead—Denmark has already announced its intention of going completely cashless. There are plenty of reasons to

The Chemistry of Money
By Brian Rohrig
© Brian Rohrig 2021
Published by the Royal Society of Chemistry, www.rsc.org

eliminate circulating currency, with convenience and crime (less of it) being the two most frequently cited reasons.

However, physical money is more convenient if you need to pay the babysitter or give your grandchild a birthday gift. When e-readers came out many predicted the end of print matter, but even millennials enjoy the feel of a real book in their hands. Therefore, writing the obituary for physical money seems a bit premature. Statistics confirm this sentiment—both the US and the UK have seen a steady rise in circulating currency in recent years. The number of euros in circulation has also continued to grow each year, and world-wide the amount of circulating cash is increasing by 5–8% annually.[1] All signs point to coins and banknotes continuing to be a fixture in society.

7.1 CREDIT CARDS

Physical money has faced threats before, with perhaps credit cards posing the biggest one, yet today they coexist in a state of happy equilibrium. The concept of credit is nothing new, but credit cards didn't come on the scene until the 20th century. Their precursor was charge coins, which were issued by merchants and valid only in their store. Some of the earliest charge coins, dating back to around 1865, were comprised of celluloid, the first thermoplastic, while others were made of metals like copper, aluminum or steel.

Metal charge plates, typically made from aluminum, were the precursor to the modern-day credit card, being issued by department stores and oil companies, again specific to only a single establishment. Often referred to as Charga-Plates, they looked like dog tags, with the customer's name and address embossed into the metal.

Credit cards that could be used in more than one location arose in the 1950s, and were initially made of paperboard. One of the first was the Diner's Club card, which came out in New York in 1950. It wasn't long before these paper cards were replaced by something more permanent. The first plastic credit card was issued by American Express in 1958, coinciding with the explosive growth of synthetic polymers in the first half of the 20th century. Nothing evoked the zeitgeist of the space age like plastic. Tupperware, Saran Wrap, and Hula Hoops all attested to plastic's versatility, and fun.

Plastic credit cards are typically made from polyvinyl chloride acetate (PVCA), a copolymer made from vinyl chloride and vinyl acetate. Copolymers are made of two different monomers, which react to form a polymer composed of differing units within its chain. About 90% of the monomers in PVCA are vinyl chloride, making its properties similar to PVC, a common component in plastic pipes. A copolymer will exhibit different properties than if the same two polymers were simply mixed together. Plasticizers and dyes are also added to the plastic during its manufacture. The credit card itself is a laminate made of three separate sheets of plastic—the color and graphics appear on the core, over which is laid two clear sheets joined together by heat and pressure.

Even though the plastic used in credit cards is incredibly resilient, the latest fad is the metal credit card, with most of the major credit card companies issuing some type of metallic card. You can get gold, platinum and titanium cards. The names can be misleading, however; the gold card is only gold-plated, while platinum cards are made from stainless steel. The American Express Centurion card is made from anodized titanium, but Mastercard's Titanium card is made from stainless steel. Others are made of proprietary blends of metals. Most of these metal cards do contain metal, but usually as a core sandwiched between two layers of plastic.

The metal credit card is seen as a status symbol, as most people can't afford to get one—the annual fee can be quite steep. If you are privileged enough to receive one it does come with the added bonus of getting your signature laser engraved onto the card. There is no practical reason to include metal in a credit card—it is purely done as a marketing ploy. A common complaint among metal card holders is that they are just too blasted heavy, as well as being inflexible. The heaviest of all is the palladium card issued by J. P. Morgan Chase, weighing in at a whopping 27 grams, being offered only to their most elite customers. It is not pure palladium, however, but it is palladium plated. It also contains a little 23-K gold. As the dimensions of a standard credit card are $85.6 \times 53.98 \times 0.76$ mm (with a volume of 3.51 cm^3) a card made of pure palladium would have a mass of

42 grams, and would cost nearly $2000 for just the metal, no doubt making the annual fees even higher.

These metal cards are just an anomaly, however, representing only the tiniest sliver in a pie that is overwhelmingly plastic. In terms of durability, corrosion resistance and hydrophobicity, nothing beats plastic, which is also flexible and lightweight. Credit cards and their kin provide a useful parallel to money—in just 100 years they have made the journey from metal to paper to plastic. Perhaps it is inevitable, then, for money to follow the same path, forcing a redefinition of what it means to pay with plastic.

7.2   TYVEK MONEY

As banknotes have become increasingly high-tech, it was only a matter of time before the substrate itself would be viewed as terribly old-fashioned and in need of a makeover. In a way, credit cards provided the first test run for plastic money, experiencing every bit as much abuse as money, and maybe even more if you consider the knuckle-busting machines credit cards used to have to go through every time a transaction was made. As credit cards are virtually indestructible, making money out of the same thing might not be such a bad idea.

The first plastic money was produced by the American Bank-note Company for the country of Haiti, appearing in 1980 (Figure 7.1). These notes were made from Tyvek, an extremely

Figure 7.1   1-gourde Haitian banknote.
© Shutterstock.

durable form of high-density polyethylene (HDPE) manufactured by Dupont. These Tyvek notes looked and felt like cotton-linen banknotes, an important consideration if they were expected to serve as their replacement. Cotton-linen banknotes themselves were only able to replace their predecessor because they looked and felt so much like paper.

There's a good chance you've seen Tyvek and not even known it. If you've ever gone to a venue where you were issued a colored paper-like wristband, then you've encountered Tyvek firsthand. If you had trouble removing it at the end of the day then you know how tough it can be—it is almost impossible to tear. The bibs you wear in a 5-K race are made of Tyvek, as is the white wrap that is used as a protective moisture barrier in new homes. Those really tough white shipping envelopes, protective DVD sleeves, and disposable cleanroom suits? All Tyvek. Tyvek is extremely durable, virtually puncture-proof, and offers great stain and fade resistance. It provides a great microbiological barrier as well, yet is permeable to gases, making it suitable for a variety of medical applications.

Tyvek is a unique type of fabric—it is non-woven, being comprised of very fine HDPE filaments that have been compressed to make a dense mat. Under magnification it resembles the thick mat of cellulose fibers you might see in paper, making it easy to see why it does such a superb job of imitating it.

Tyvek might have had a long and healthy relationship with the banknote world if not for one fatal flaw—it was difficult to get the ink to adhere, especially in the humid climate of Haiti. Sadly, the ink was susceptible to smudging, rendering the banknotes almost unreadable after a few years. Therefore, Tyvek's legendary stain resistance proved to be its undoing, a problem that might have had a solution were it not for there being so many other options available—there are no shortage of polymers to choose from, with new ones constantly being synthesized.

7.3 POLYPROPYLENE

Australia is usually credited with introducing the first successful polymer banknote, with the issuance of a $10 note in 1988 commemorating its bicentennial. The choice of a commemorative note was a calculated move to gauge the note's success without breaking the bank—commemorative notes have limited circulation and can be withdrawn without anyone noticing, as it is not the mainstay for that particular denomination; the cotton-linen $10 notes were still being issued. These polymer notes didn't happen overnight, being the product of 20 years of careful research in a collaboration between the Commonwealth Scientific and Research Organization (CSIRO) and the Reserve Bank of Australia. The initial impetus for the polymer note was the high rate of counterfeiting of the existing $10 note. Even though it was made with all the latest security features, the market was flooded with counterfeits within a year after its issuance. The goal was not only to change the substrate but also to incorporate the latest security features—a great deal of work was done on diffraction gratings, moiré patterns and other bells and whistles, with the express goal of creating a banknote that could not be counterfeited.

After much research, it became evident that polypropylene would be the best material from which to make the substrate. From the beginning, it was called a polymer banknote, so as to avoid the negative connotations associated with plastic, which is often used as a synonym for being cheap or of poor quality.

Polypropylene is similar to the polyethylene used in Tyvek—both are classified as polyalkenes; in industry all fibers made from these polymers are referred to as olefins, a term based on the Latin word *oleum* for oil. Olefins do have an affinity for oil, a weakness that is evident in untreated fibers.

An alkene is an unsaturated hydrocarbon, containing at least one double bond between the carbon atoms. Propylene, also known as propene, has a formula of C_3H_6; it is similar to propane, the compressed gas used for propane torches, but propene is more volatile than propane, burning with a hotter flame. This greater reactivity is a result of the double bond between two of propylene's carbon atoms, which by itself is stronger than a single bond; however, every double bond contains both a pi bond and a sigma bond (Figure 7.2). The sigma bond occurs between the s orbitals, and the pi bond is between the p orbitals; there is considerably more overlap between the orbitals in a sigma bond, thus it is more stable than a pi bond. This weaker pi bond plays an important role in the polymerization of the alkenes.

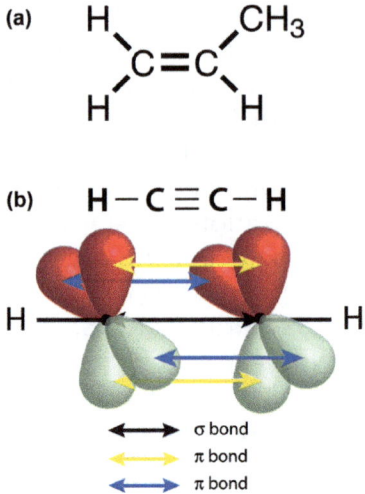

Figure 7.2   (a) The structure of propylene, and (b) the sigma and pi bonding.

It's no accident that the most commonly synthesized and most widely used plastics on our planet are all made from alkenes, which are identified by their -ene ending. Ethylene, styrene, propylene, and so forth will undergo polymerization to become polyethylene, polystyrene, polypropylene, and so on. Each of these polymers are formed by addition polymerization—a straightforward linking together of the monomers with nothing added or lost in the process, unlike the formation of cellulose, which as you may recall forms as the result of a condensation reaction in which a water molecule is released whenever two glucose molecules join up.

Alkenes are susceptible to polymerization because of the weak pi bond, which can be attacked by any of a variety of reactants—when it is broken a highly reactive radical can form that readily reacts with another monomer. These monomers are added one at a time to the end of the chain, ending in the linking together of thousands of these subunits, resulting in a very stable structure. Other mechanisms to form polyalkenes involve the use of a catalyst; regardless of the method used, the weak double bond of an alkene is always exploited, ending with the formation of a much more stable polymer composed of all single bonds.

If the methyl groups in polypropylene are randomly arranged, appearing on opposite sides of the chain, then the atactic variety of propylene is formed. This form of propylene was the first type synthesized, but as it tended to be soft and gooey, its usefulness was limited. However, if the methyl groups all end up on the same side, then isotactic polypropylene is formed, which is stronger and harder, having a much higher melting point (around 160–170 °C). This form of polypropylene was first synthesized in 1953 by Karl Ziegler and Giulio Natta, the discovery of which propelled them to the Nobel Prize in Chemistry in 1963. The predominant method used to make polypropylene today involves the Ziegler–Natta catalyst.

Most products that utilize polypropylene will be of the isotactic variety (Figure 7.3). These products are known for their durability—the living hinges on prescription drug bottles are made from isotactic polypropylene, offering excellent fatigue resistance. Polypropylene can be crystal clear or can be made to be completely opaque, as it accepts pigments very well. It is used for a variety of containers, such as shampoo bottles and the like, as well as disposable cups that are not only resilient, but soft and supple—to find out if something is made from polypropylene look for recycling code #5. Polypropylene has a pleasant slippery feel to it. It is easy to tell if a piece of plasticware is made of polypropylene—you can bend it and it won't break; polystyrene, on the other hand, is more brittle. Polypropylene is a great choice for food storage containers as it tends to be stain-resistant, and its relatively high melting point allows it to be placed in the dishwasher.

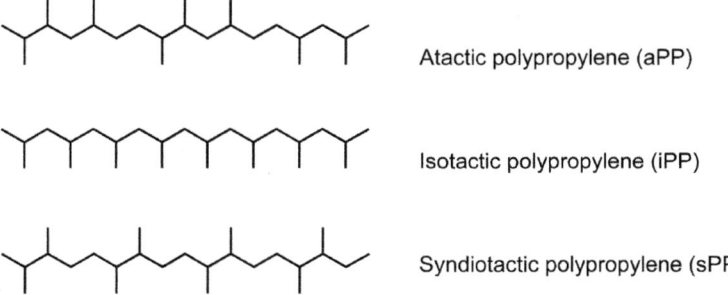

Atactic polypropylene (aPP)

Isotactic polypropylene (iPP)

Syndiotactic polypropylene (sPP)

Figure 7.3   Polypropylene structures.

7.4   PLASTIC MONEY

To make a banknote out of plastic it must first be made into a film, which is defined as any plastic with a thickness less than 0.25 mm; anything thicker is referred to as sheeting. A typical polymer banknote will have a thickness less than 0.1 mm. Virtually any type of plastic can be made into a film, which is easy to demonstrate. If you have a heat gun, simply apply it to the side of a plastic milk jug (typically HDPE) until an area the size of your palm becomes soft and transparent. If you then blow into the jug the plastic will expand into an enormous bubble.

To create commercial-grade polymer film a bubble is also formed, but this bubble is about 5 stories high. The process begins when plastic pellets are melted and then poured into a tube, from which they are then extruded through a small opening in an annular (ring-shaped) die through which air is blowing. A giant bubble is created, which is cut and collected into rolls as it cools. The blown film method is one of the more common ways polymer films are made. As the film forms it is stretched in two directions—both in the longitudinal and transverse directions. The resulting product is referred to as BOPP (biaxially oriented polypropylene), which is known for its high tensile strength, clarity and chemical resistance.

If you hold a polymer banknote up to the light you will see that it is mostly opaque—before any printing occurs it is coated with several layers of thick white ink, providing a surface to which printing ink is more likely to adhere. The only area not inked is the clear security window, which often has an additional security feature such as a hologram or diffraction grating applied over it. After printing and the application of the security features, a protective coat of polyurethane varnish is applied over the surface of the banknote. To prevent the surface from being too slippery, finely ground silica (SiO_2)—the primary component of sand—is added, keeping the bills from sticking together.

The first trial bills produced in Australia were $3 and $7 bills, which served as a precaution to keep them from escaping into circulation. As there was little precedent for plastic money, these first trial notes underwent a series of grueling tests to make sure they were up to the task. They were stained with all variety of food, drink and household cleaners, undergoing rigorous tests involving abrasion, crumpling, folding, UV exposure, and biological factors. They were subjected to high altitudes and to the depths of the sea, and as there is no shortage of inhospitable terrain in Australia, the perfect laboratory in which to test the bills was all around. The last thing the Reserve Bank wanted was for these notes to begin circulation and then face a catastrophic failure owing to some unforeseen factor, such as happened in Haiti.

When the polymer banknotes finally made their appearance, they were well-received by the public, effortlessly blending in with existing currency—they looked and felt like just the old cotton-linen banknotes, and even worked in ATMs. Within a decade, Australia would convert completely to plastic, one of the first countries to do so. A number of countries have since introduced plastic currency, including Canada, Mexico, and the UK. Most countries that have tried plastic banknotes have eased them in gradually, though, keeping the no-plastic notes in circulation in case the plastic ones don't work out.

The polymer bills have been a smashing success—one of the biggest benefits has been a drastic reduction in counterfeiting. The plastic substrate allows for the incorporation of security features not possible on paper. The clear security window is perhaps the hardest feature to replicate, which hasn't stopped some from trying. One confiscated counterfeit contained a piece of clear plastic taped over a hole. Others have tried printing fake bills on waxy paper, or have applied a plastic coating to a counterfeit to simulate the look and feel of the polymer notes. Perhaps the best way to distinguish between a paper and plastic note is to attempt to tear it in half. A plastic note just will not tear. Once a note does start to tear, however, it tends to keep tearing. Even though they cost more to make, the plastic notes last considerably longer—countries that have made the switch to plastic have noticed a substantial overall savings in the cost of banknote production.

7.5 SHRINKY DINKS

One common complaint about the plastic banknote is that it shrinks when heated, with reports of the notes being ruined if ironed or placed in the microwave. I had a 10-naira Nigerian banknote I decided to sacrifice to test this contention. (It wasn't a huge sacrifice, however, since the exchange rate for 1 naira is only 0.0028 US dollars.) The microwave didn't affect it but a toaster oven most certainly did, causing it to quickly shrivel up into a sad wad of plastic.

The banknote behaved the same way as the plastics in the aptly named Shrinky Dinks craft kits for kids. These kits contain sheets of polystyrene, the surface of which has been roughened up so you can draw on it. When the shrunken version emerges from the oven it is perfectly proportioned, albeit much smaller. Shrinky Dinks only work because of how the plastic is manufactured. First, the molten polystyrene plastic is extruded through a slot die and rolled into a thin sheet. It is then stretched and pulled in all directions, becoming biaxially oriented. When it cools, the polymer chains are frozen in this elongated, high energy state, like stretched-out springs. When the plastic is heated and softens, the polymer chains revert back to their initial, more relaxed, lower-energy state, causing them to shrink back to their original size. If a piece of plastic is initially pulled in only one direction while being formed, however, then heating will cause it to shrink only in the direction in which it was originally stretched.

7.6   THE ANIMAL FAT BROUHAHA

Not all countries with plastic banknotes utilized Australia's recipe. Canada's polymer notes are made out of polyethylene terephthalate (PET), the same ingredient used to make soda bottles. In 2016 the Bank of England began issuing £5 polymer notes, followed by £10 polymer notes the following year. These notes were made from polypropylene, but that's not all they were made from—in 2016 it was revealed that they contained traces of tallow, a product made by rendering animal fat, primarily beef, raising an outcry from vegans and those whose religious beliefs preclude them from touching animal products. Similar concerns were also raised in Canada and Australia.

Tallow is easy to make—just brown some ground beef or bacon in a frying pan and pour off the liquid fat into a jar; when it cools and hardens you have tallow, provided only fat made it into the jar and it was heated long enough to drive out all traces of moisture. Tallow is often used to make soap and candles, but can also be used for cooking. Tallow contains a lot of saturated fats and has a long shelf life. It is added to molten polymer mixes as a lubricant, keeping the molten plastic flowing freely, as well as preventing it from sticking to metal surfaces.[2] In response to criticism over these animal products in their banknotes, the Bank of England pledged to work with their suppliers to consider plant-based substitutes, such as palm oil, which might accomplish the same purpose as tallow, but with a lot less controversy.

7.7 POLYMER COINS

Banknotes aren't the only type of currency to utilize polymers—polymers are also being incorporated into coins. In 2016 Germany issued the world's first coins containing a polymer. These €5 bimetallic coins contained a translucent blue plastic ring embedded within the coin, separating the outer ring from the inner disk. As the polymer ring is inserted before the coin is struck, the plastic must be extremely tough, being able to endure the rigors of the minting process without losing its thermoplastic properties. It must then be able to withstand all the wear and tear a normal circulating coin endures. The special type of proprietary plastic used required eight long years of research in a cooperative effort between the Baden-Wuerttemberg Mint, the Bavaria Mint, and the DWI Leibniz Institute for Interactive Materials at the RWTH Aachen University. It is legal tender only in the Federal Republic of Germany.

The center disk is made from 75% copper and 25% nickel, while the outer ring is composed of 81% copper and 19% nickel. These handsome silver-colored coins contain an image of the world on the obverse side and an eagle on the reverse (Figure 7.4). The blue ring is designed to represent the atmosphere, which surrounds the earth. The polymer ring can be injected with any color, including fluorescent pigments. Germany has also released a collector's coin with a red polymer ring as part of their climate zones of the earth series—the first coin minted featured the tropical zone. The melding of metal and plastic not only makes this coin unique, but also makes it almost impossible to counterfeit.

Germany is not the only country to use plastic in their coins—the country of Transnistria has issued pure plastic coins, doing

Figure 7.4   Germany five euro coin with planets and a blue polymer ring.
© Shutterstock.

away with metal altogether on their 1, 3, 5 and 10-ruble coins. Although the coins were intended to replace banknotes of the same denomination, they actually blur the distinction between the two—it is difficult to know just exactly what they are.

Each coin of a different denomination is a different shape and color. The yellow 1-ruble coin is circular shaped, the green 3-ruble coin is a square, while the blue 5-ruble coin is shaped like a pentagon; the red 10-ruble coin is in the shape of a hexagon. Each coin is a work of art, though, containing security features that more closely resemble banknotes than coins—each contains microtext as well as ultraviolet features.[3]

Although the UN does not officially recognize Transnistria, it has declared its independence from Moldova, an Eastern European country next to Ukraine, which was formerly part of the Soviet Union. Whether or not Transnistria can be classified as a country, and therefore issue currency, is debatable. Regardless, plastic coins, either in part or in whole, are still very much an anomaly in the coin world, and even though plastic is making significant in-roads into the banknote market, whether or not they will ever replace metal coins to any appreciable extent remains to be seen. Plastics though do possess the singular property that has defined money from the start—they are for the most part chemically inert.

7.8   THE NEW £1 COINS

Plastic coins are not the only type of money to utilize technology normally reserved for banknotes—coins are increasingly appearing with many of these same security features. One of the more exciting coins debuted in 2017—the British £1 coin, which has 12 sides, making it a dodecagon. These coins benefit from some of the most advanced coin-making technology ever used, primarily in an attempt to limit counterfeiting. One in 30 of the former £1 coins in circulation were believed to be counterfeit, necessitating the removal of two million of them from circulation each year. Like many newer coins, it is bimetallic, composed of an inner nickel-plated silver colored inner disc surrounded by an outer ring of nickel-brass alloy. The edge alternates between ridged and smooth sections, making it easy to identify by touch.

There are a number of anti-counterfeiting measures incorporated into the coin—in tiny lettering near the outer edge is the inscription "ONE POUND" repeated over and over, while on the reverse the date appears in several places along the rim. This tiny lettering is inscribed by a high-tech cutting tool utilizing a laser, which would be extremely difficult for a counterfeiter to replicate. A shape-shifting image on the obverse of the coin looks like a "1" when held one way and a "£" symbol when slightly turned (Figure 7.5).

The seemingly endless number of intricate security features has enabled the Royal Mint to boast that the new £1 coin is the most secure coin in the world. One of the most intriguing features involves the use of iSIS technology, which stands for Integrated Secure Identification System. According to Stuart Wilson, Head of Commercial Development at the Royal Mint, tiny particles are embedded within the plated layer of the coin, which is much more effective than simply applying them to the surface. The iSIS particles are incorporated within the plating mixture as

Figure 7.5   The new British £1 coin.
© Shutterstock.

it is being electrodeposited onto the coin, ensuring their even distribution throughout the entire layer. On average, coins wear down 1 μm per year, so anything applied to the surface would soon be gone. As the outside plating of the coin is 25 μm thick, these embedded particles will stay put for the lifetime of the coin.

When the coin is placed under an appropriate illumination source, a coin embedded with the iSIS security system will emit energy of a specific wavelength that is picked up by a special optical detector, verifying the authenticity of the coin. Unlike other detection devices that work within an acceptable range, the iSIS detector provides a simple yes-or-no answer—either the coin is genuine, or not. The particles are tiny, but they are visible using a scanning electron microscope under a magnification of 3000×. The Royal Mint is not releasing the identity of these particles—they will only say they are optically active. According to Wilson, the iSIS offers banknote level security for a coin, making the new £1 coin very difficult to counterfeit.[4]

7.9 FLUORESCENT COINS

The £1 coin is not the first to incorporate ultraviolet security features. In 2014 the Cook Islands, in conjunction with the Royal Mint, issued a stunning one-dollar commemorative coin containing a full-color dinosaur, featuring either a tyrannosaurus rex, diplodocus, stegosaurus, or triceratops. Under a blacklight, the skeleton of the dinosaur appears, along with its name. The following year a five-dollar coin containing a scene of five dinosaurs was issued—when exposed to UV light, the skeletons of all five dinosaurs appear.

In 2015, the Royal Canadian Mint issued a set of four pure silver coins, each highlighting a different weather phenomenon. The first coin in the series featured a lightning bolt that fluoresces under a blacklight. The tiny island country of Niue, near New Zealand, working with the Mint of Poland, has issued a beautiful "Code of the Future" series of coins that each fluoresce under a blacklight. The first coin featured a head containing computer circuits holding the earth, and the next celebrates the speed of light.

Even more exciting than fluorescent coins are glow-in-the-dark coins. One of the first glow-in-the-dark coins to enter general circulation was the 2017 Canadian toonie, issued to celebrate the country's sesquicentennial. The coin featured two canoeists rowing under the glow of the aurora borealis, or northern lights. These tri-colored coins featured a golden lake, blue sky, and green streaks representing the northern lights. If placed in a dark room, the green streaks representing the northern lights glow. In 2017 Canada issued another glow in the dark coin—a pure silver coin plated with a layer of black ruthenium. In the center of the coin is a UFO within a maple leaf. The UFO, fittingly, glows in the dark. Ruthenium makes for a fine finish, as it is very hard and extremely corrosion resistant; it is in the platinum group, and therefore shares platinum's non-reactivity. A look at an older periodic table will reveal that platinum and ruthenium are in the same group—group VII.

7.10   THE FUTURE OF MONEY

If you were to add up all the money in the world—the cash and coins—you would have around $5 trillion dollars,[5] a trifling sum when you consider that all the assets of every person and business in the world would total $90 trillion dollars, leading to the disturbing conclusion that there is a lot more virtual money in the world than physical money. Armed with this knowledge, we can throw back the curtain and reveal that most of today's money is nothing more than a few bits of data stored on a computer.

A bit, short for binary digit, is the smallest amount of information that can be stored on a computer, and in the current binary system, is represented by a 0 or 1, which is the most efficient way to store data, with 8 bits equal to 1 byte. If I had $1000 in the bank, it would be stored as 1111101000. All the money in the world—all $5 trillion of it—in binary code would be represented as 1001000110000100111001110010101000000000000. In 2018, the worldwide storage capacity of all the word's data centers stood at 1.45 zettabytes (1 zettabyte $= 1 \times 10^{21}$, or 1 sextillion, bytes), plenty of room to store the world's money.

Text is stored the same way. Most computers use the ASCII (American Standard Code for Information Interchange) coding system, in which each letter is given an 8-number code. The letter A, for example, is represented by 01000001, B is 01000010, and so forth.

There have been numerous ways to store this data. The old-fashioned floppy discs contained tiny little iron oxide particles embedded into a flimsy plastic base. These particles, like magnets, could be flipped one way or the other—for a 0 a particle would be oriented one way, and for a 1 it would be oriented the other way. As you only need a physical medium that can exist in two separate states, magnets are perfectly suited to record binary data—they can easily be switched from one orientation to another.

We often hear about our data being stored in the cloud, which of course means it is not stored on our computer but someplace external, but it's still physically stored somewhere, often in much the same way as it might be on our own personal devices. The difference is that cloud data is stored in the hard drives of servers, which are housed in server farms (or data centers) where racks of them are housed, often numbering into the thousands, or in the case of Google, the millions.

A magnetic hard drive is composed of a circular disk known as a platter, which is made of a durable non-magnetic material such as glass, ceramic or some kind of specialized alloy (Figure 7.6). This disk is coated with a thin layer of tiny magnetic grains about 10 nm in diameter—CoPtCr is one alloy commonly used. As the read–write head passes over the hard drive, a number of these tiny grains will line up in the same direction, forming a magnetic

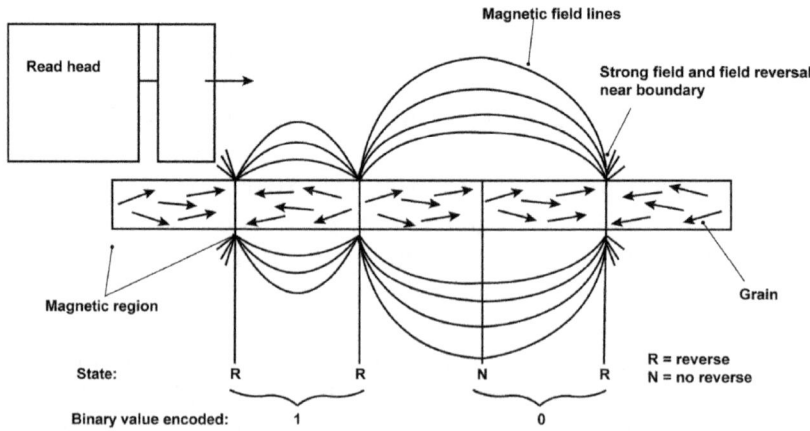

Figure 7.6 Magnetic hard drive. Original image source: Yale University.

domain. This domain represents one bit of data, perhaps a 0. If the grains line up in the opposite direction, they create a 1. A magnetic hard drive works in basically the same way as a floppy disk, it's just that the substrate is harder and the particles are a lot smaller, making them capable of holding a lot more information on a more permanent basis.

Although magnetic hard drives still have their place, they are being replaced by solid state drives, which have no moving parts. They still store data in binary form, except there are no magnets involved, but rather only electrons, which are used to charge transistors—a charged transistor represents a 1, while an un-charged transistor translates into a 0.

The transistor has been called the most important invention of the 20th century, and by replacing bulky vacuum tubes they enabled electronic devices from computers to radios to become much smaller. Transistors are typically made from silicon, or some other semiconductor, and are doped with elements such as boron and phosphorous to increase their conductivity.

In the simplest terms, a transistor is a device that regulates the flow of current, and typically contains three electrodes. It can amplify current, or it can be used as a switch. Transistors are tiny—your smartphone contains billions of them, and your laptop billions more (Figure 7.7). They keep getting smaller—some transistors are only 1 nm long! Solid state storage relies on transistors to store data—each transistor acts like a switch that can be turned on or off. If voltage exists across a transistor then current will flow and the transistor is on—which equates to a 1; if there is no voltage and no current flow then the transistor is off—which equates to a 0. Therefore, if I wanted to store the letter A (01000001), I could do it with eight transistors in a row, arranged as follows: off-on-off-off-off-off-off-on.

This scenario is, of course, highly simplified, and this most basic of explanations was only offered up to demonstrate what your bank account looks like in the cloud. From a computer's perspective, you are just a few electrons away from being either a pauper or a millionaire.

Regardless of the form your money takes, there will always be a physical manifestation of it somewhere; after all, electrons do have mass. Even cryptocurrencies such as the Bitcoin would not exist without physical computers to solve problems and perform

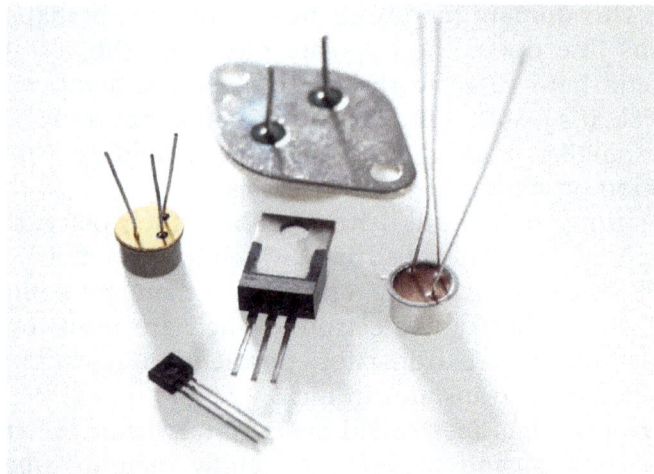

Figure 7.7   Various types of transistors, nano-sized transistors, and a micro-chip containing numerous transistors.
© Tek Image/Science Photo Library.

transactions and store data. Even if all tangible money does end up being eliminated, we will still need ways to transfer it. Instead of metals and plastics going into making money, these materials will instead go into the devices we use to access that money, be it a phone, a card or even a chip implanted beneath the skin. From the beginning money and chemistry have been inexorably intertwined in a marriage that is still healthy and strong, a union that will likely continue well into the foreseeable future.

REFERENCES

1. Currency Research, Cash in circulation continues to grow by 5% annually, 2017, https://www.cashmatters.org/blog/cash-circulation-growing-5-annually/.
2. R. Lowe, Tallow in plastics – why? Impact solutions, 2016, https://www.impact-solutions.co.uk/tallow-in-plastics/.
3. P. Saha, Transnistria 2014 – New coin family in synthetic material, World Coin News, 2014, http://worldcoinnews. blogspot.com/2014/08/transnistria-2014-new-coin-family-in. html.

4. S. Wilson, Introducing an effective coin system, Talk given at the 10th Technical Forum, 43rd World Money Fair, Berlin, Germany, 2014, https://coinweek.com/featured-news/world-mints-10th-technical-forum-presentations-berlin-germany/.
5. M. Hartman, Here's how much money is in the world—and why you've never heard the exact number, Business Insider, 2017, https://www.businessinsider.com/heres-how-much-money-there-is-in-the-world-2017-10.

Subject Index